T. REX

W9-AAG-303

AND THE CRATER
OF DOOM

Walter Alvarez

T. REX
AND THE CRATER
OF DOOM

Walter Alvarez is a geophysicist who
teaches at the University of California
at Berkeley and is a member of the
National Academy of Sciences. He lives
in Berkeley, California.

T. REX

AND THE CRATER

OF DOOM

T. REX
AND THE CRATER
OF DOOM

Walter Alvarez

Vintage Books
A Division of Random House, Inc.
New York

FIRST VINTAGE BOOKS EDITION, AUGUST 1998

Library of Congress Cataloging-in-Publication Data
Alvarez, Walter, 1940–
T. rex and the crater of doom / Walter Alvarez.
Originally published: Princeton, N.J. :
Princeton University Press, c1997.
Includes bibliographical references and index.
ISBN 0-375-70210-5
1. Catastrophes (Geology). 2. Extinction (Biology).
3. Cryptoexplosion structures—Mexico—Campeche,
Bay of, Region. 4. Chicxulub Crater. I. Title.
[QE506.A48   1998]
576.8′4—dc21   98-9369
CIP

Author photograph © Robert Holmgren

Random House Web address: www.randomhouse.com

Printed in the United States of America
10 9 8 7 6

This book is affectionately dedicated to Milly Alvarez;
a skillful and compassionate leader in her own field
of mental health, and the perfect companion during
thirty years of geological expeditions on five continents

CONTENTS

This is the story of one terrible day in the history of the Earth.

We come into this world in ignorance of everything that happened before we were born. As we grow up, we learn about the events of bygone years. From parents and relatives we hear the history of our families, and in books we read the history of humankind. But the past of the planet on which we live extends back through vast stretches of time before human beings appeared, and no eyewitnesses, no firsthand accounts can tell us about the history of the Earth.

And yet, we can learn a great deal about Earth history, because that history is recorded in rocks. Geologists and paleontologists are the Earth historians who read that record—examining rock outcrops in remote parts of the world and bringing back samples for analysis in the laboratory. By observing and measuring and interpreting the information held fast through the eons in solid rock, geologists and paleontologists have pieced together an understanding of the history of our planet, back to its beginning 4,600 million years ago.

What kind of past has it been? Is Earth history a chronicle of upheavals, catastrophes, and violence? Or has our planet seen only a stately procession of quiet, gradual changes? Most early students of the Earth were catastrophists, but as geology matured as a science, geologists found that Earth features, even dramatic ones like the Alps and the Grand Canyon, can best be explained by slow, gradual changes acting over the enormous time spans of Earth history. Geologists embraced gradualism as the proper explanation for everything they saw in the rock record and came to shun the notion of great catastrophes in the past.

Today a less restricted view is emerging, blending gradualism and catastrophism, which are no longer seen as mutually exclusive. Geologists continue to find that *most* changes in Earth history have taken place slowly and gradually. But there is a new recognition that on a few occasions the Earth has suffered enormous catastrophes, which have totally redirected the subsequent course of events.

This book is the story of how Earth historians uncovered the evidence for one great catastrophe in the Earth's past—the impact of a huge rock that fell from outer space 65 million years ago, excavating an enormous crater in the Yucatán Peninsula of Mexico and causing such disturbance to the environment that a wide variety of plants and animals perished forever. The most famous of the victims in this cataclysm was the great carnivorous dinosaur, *Tyrannosaurus rex*.

The story of the impact and the extinction it caused is dramatic and horrifying, although the passage of 65 million years makes it seem comfortably remote. But running parallel to the historical account of the event itself is the very human story of how Earth historians discovered the rock record of the great impact and learned to read and interpret the evidence. It is a story of challenge by a few geologists to what their science had long believed to be true, of staunch defense of the traditional view by other geologists, of conflict and friendship, of adventure in distant places, of painstaking measurements in the laboratory, of mystification and discovery, and of the common effort by scientists from many countries to solve an absorbing mystery.

It is also the story of how geology and the other disciplines which study the Earth have emerged as fully mature sciences, distinguished by their inherently interdisciplinary nature, by the complexity of their subject matter, and by the obvious requirement to move from reductionistic to holistic science in order to achieve their central goal of understanding the Earth. Through the twentieth century, physics and chemistry, and recently molecular biology, have made enormous strides in understanding Nature by the analytical approach—by reducing problems to their fundamental components and studying these components in isolation. In the twenty-first century, science will be in a position to begin putting the pieces together, in order to seek a synthetic or holistic understanding of Nature. The Earth sciences are inherently synthetic and are therefore uniquely placed to lead this development. The story of research on impacts and mass extinctions illustrates in detail how this can happen.

It has been my privilege to be involved in this adventure from the first discovery of evidence for a great impact at the time of the dinosaur extinction. For the first ten years or so, many scientists found more and more evidence supporting that impact, but all efforts to find the crater were in vain. Finally, in 1991, the long-sought crater was recognized, buried beneath the Yucatán Peninsula.

Much research remains to be done before we reach a full understanding of the catastrophic impact 65 million years ago. But the recognition of the crater was a turning point, so this is an appropriate time to tell the story of its discovery. In doing so, I have tried to make this book accessible to anyone with a general interest in science. In addition, by including extensive notes and references, I have tried to make it useful as a starting point for those wishing to delve more deeply into these matters.

Anyone reading this book will recognize my indebtedness to many scientists. Among them I would single out Luis W. Alvarez, Frank Asaro, Alfred G. Fischer, William Lowrie, Richard A. Muller, Eugene M. Shoemaker, and Jan Smit, along with my students and postdoctoral researchers over the years who make up the Renaissance Geology Group.

Several people have helped me greatly with their comments on the manuscript, especially Milly Alvarez, Frank Asaro, Carol Christ, Philippe Claeys, Dan Karner, Rudy Saltzer, and Gene Shoemaker. The toughest and most detailed critiques came from Rich Muller, who went beyond the obligations of friendship in helping me to improve the structure of the book and to polish the rough spots. My artist friend Vincent Perez painted the dust jacket on the basis of careful discussions of what a cataclysmic impact might look like to a dinosaur unfortunate enough to see one close up.

It has been a pleasure working with Princeton University Press. Faculty Advisor Alfred G. Fischer and former Science

Editor Edward Tenner, now a best-selling author in his own right, began talking with me about doing a book many years ago. Current Science Editor Jack Repcheck finally persuaded me to write the story of the great impact, and skillfully shepherded the project to conclusion.

T. rex and the Crater of Doom comes with warm thanks to three groups of people: First to my scientific colleagues on all sides of the extinction debate, all over the world, who have made this the most exciting intellectual adventure I can imagine. Then to the citizens of the State of California, who employ me at their splendid university to teach geology to their sons and daughters. And finally to the people of my country, who support the American research enterprise with their taxes, through agencies like the National Science Foundation and NASA. I hope they take pleasure from this story of discovery.

T. REX

AND THE CRATER

OF DOOM

Armageddon

But it was too late. At that moment the rock quivered and trembled beneath them. The great rumbling noise, louder than ever before, rolled in the ground and echoed in the mountains. Then with searing suddenness there came a great red flash. Far beyond the eastern mountains it leapt into the sky and splashed the lowering clouds with crimson. In that valley of shadow and cold deathly light it seemed unbearably violent and fierce. Peaks of stone and ridges like notched knives sprang out in staring black against the uprushing flame in Golgoroth. Then came a great crack of thunder.

—J.R.R. TOLKIEN, *The Lord of the Rings*

REQUIEM FOR A LOST WORLD

Try to imagine a different world—different from the one we live in. Not wildly different, like the settings of science fiction stories which take place on airless planets or in giant spaceships. We are looking for a world much like our own, but different in subtle ways. J.R.R. Tolkien described such a world in *The Lord of the Rings*—with mountains, swamps, and plains like ours, but with a slightly different geography—much like Europe, but not quite the same. Tolkien's "Third Age of Middle Earth" has familiar inhabitants like humans and horses, but other creatures that we know well, like dogs and cats, are missing. Middle Earth also has unfamiliar inhabitants—dwarves, elves, wizards, and hobbits. It is terrorized by the merciless, sharp-clawed goblins called orcs. Tolkien's world seems ancestral, or perhaps alternative to ours.

The world we seek is reminiscent of Tolkien's Middle Earth. It has mountains, deserts, forests, and oceans, arranged in a geography that is something like our Earth, yet noticeably different. It has rivers and canyons, plateaus and sand dunes. It has cloudbursts in the mountains, and glowing sunsets in the clear air after a thunderstorm. Some of the inhabitants seem familiar, though not exactly like the ones we know. Evergreen trees and deciduous trees shade the landscape, and the streams are full of fish. But the ground is bare of grass, and the animals look different. Little furry ones are recognizable as mammals. But there are also giant creatures, some placidly grazing while others hunt, with claws as terrifying as those of any orc in Middle Earth.

This world is different from ours, but it is familiar through museum reconstructions, paintings, and films. For this is not the Third Age of Middle Earth, but rather the Third Period of Middle Life. Geologists use the term Mesozoic, or Middle Life, for the Age of Dinosaurs. The third period of the Mesozoic was the Cretaceous, following the Triassic and the Jurassic periods.

More precisely, the world we are imagining was the very end of the Cretaceous, 65 million years ago. It was ancestral to our modern world, with a geography that was different but still familiar, because continental drift since then has moved the Earth's land masses around but has not completely rearranged them. India had not yet collided with Asia to thrust up the Himalayas, but there were already mountains in western North America. Sea level was higher than today, and part of the interior of North America was covered by a shallow sea.

Not only was that world ancestral to ours, it was in some sense alternative as well. For it was a stable world. Despite the violent hunting of the carnivorous dinosaurs and the oft-depicted dramatic battles between *Tyrannosaurus rex* and *Triceratops*, life patterns and the inhabitants themselves had changed only slowly during the previous 150 million years. The dinosaurs were very successful large animals and shared their world with equally successful small animals and with plants of all kinds. There is every reason to believe that if it had remained undisturbed, the Mesozoic world could have continued indefinitely, with the slightly evolved descendants of the dinosaurs dominating a world in which humans never appeared.

But the Mesozoic world did not remain undisturbed. It ended abruptly, and with no warning, 65 million years ago. Vast numbers of highly successful animal and plant species suddenly disappeared in a mass extinction, leaving no descendants. This break in the history of life is so impressive that geologists use it to define the boundary between the Cretaceous, or last period of the Mesozoic, and the Tertiary, or first period of the Cenozoic. Today's world is populated with descendants of the survivors of the mass extinction that ended the Cretaceous world.

Looking back across the abyss of time which separates us from the Cretaceous, we can somehow feel nostalgia for a long-lost world, one which had its own rhythm and harmony. We feel a special sadness when we think about its plants and animals, fish and birds—for most of the Cretaceous animals and plants are irretrievably lost. We can even feel some sorrow as we imagine the sun setting over a western ocean, painting the clouds with orange and red and yellow and gold, on the last evening of that world. For the Cretaceous world is gone forever, and its ending was sudden and horrible.

THE APPROACH OF DOOM

Doom was coming out of the sky, in the form of an enormous comet or asteroid—we are still not sure which it was. Probably ten kilometers across, traveling tens of kilometers a second, its energy of motion had the destructive capability of a hundred million hydrogen bombs. If an asteroid, it was an inert, crater-scarred rock, dark and sinister, invisible until the last moment before it struck. If a comet, it was a ball of dirty ice, spewing out gases boiled off by the heat of the Sun, and it announced impending doom with a shimmering head and a brilliant tail splashed across half the sky, illuminating the night, and finally visible even in the daytime as Armageddon approached. Let us think of it as a comet, remembering that perhaps it was an asteroid instead. Comets have been mistakenly interpreted by humans in times past as harbingers of doom, foretelling famine, plague, and destruction. Although no humans were there to witness the giant comet of 65 million years ago, in this case it really did portend disaster.

The solar system abounds in comets and asteroids, some even bigger than the one which was nearing Earth on that day 65 million years ago. Most asteroids remain in a belt between Mars and Jupiter, and most comets orbit the Sun far beyond distant Pluto. Occasionally, however, an asteroid has its orbit deflected by Jupiter's gravitational pull, or a comet orbit is altered by the gravitational tug of a passing star. A few of these asteroids and comets are diverted into orbits which cross that of the Earth. An impact occurs when such an object intersects the Earth's orbit just as Earth happens to be at the crossing point. This is what is going on every time you see a shooting star flashing across the night sky. Those streaks of light are due to tiny fragments of comets or asteroids burning up through friction in the Earth's atmosphere. Somewhat larger objects, the size of a fist, are too big to burn up completely in the atmosphere, but are slowed down enough to survive their impact on the Earth's surface. These objects are the meteorites displayed in museums and studied by geologists interested in extraterrestrial rocks.[1]

Large impacts can also happen, and they were frequent in the early history of the solar system, as witnessed by the ancient, crater-scarred face of the Moon. But large impacts are rare nowadays, because the debris that was abundant in the early solar system has been swept up by the planets, large Earth-crossing comets and asteroids are now rare, and Earth is a very small target. To see how small, look at Venus just after sunset, when it is the "evening star." Venus is the size of the Earth, and from our distance it is a tiny, although brilliant, dot in the sky—a very difficult target to hit.

Earth is protected, therefore, by the fact that large comets and asteroids rarely come into the inner solar system, and those that do are unlikely to hit something as small as our planet. So we can imagine the giant comet of 65 million years ago coming close to the Earth again and again, over a period of centuries or millenia, as it orbited the Sun—sometimes far from Earth, sometimes close enough to put on a spectacular display in the night sky. A set of near misses like this must take place every now and then in Earth history, but usually the comet hits the Sun or another planet, or is deflected out of the inner Solar System. In this particular case, however, there came a time when the invader's orbit intersected that of Earth just as both were approaching the intersection point. This time there would be no escape. The comet was aimed toward the southern part of North America—toward the shallow seas and coastal plains which are now the Yucatán Peninsula of Mexico.

THE MEASURES OF DESTRUCTION

It is very difficult to appreciate the impact that was about to occur, because such an extreme event is far beyond our range of experience—for which we can be most grateful! One can write down the measures of what happened—an object about 10 km in diameter[2] slammed into the Earth at a velocity of perhaps 30 km/sec.[3] But these measures only acquire meaning when we try to visualize them, or make analogies to help our understanding. How can we imagine a comet 10 km in diameter? Its cross section about matches the city of San Francisco. If it could be placed gently on the surface of the Earth it would stand higher than Mount Everest, which only reaches about 9 km above sea level. Its volume would be comparable to the volume of all the buildings in the entire United States. It was a big rock, or a big ice ball, but not of a scale beyond our comprehension.

What turned it into a cataclysmic weapon was its velocity. The estimated impact velocity of 30 km/sec is 1,000 times faster than the speed of a car on the highway and 150 times faster than a jet airliner. It is about 6 times faster than the speed of seismic waves in rock. When a collision takes place at velocities this high, our experience is not a useful guide, and rock materials do not behave in the ways we are used to. Instead, a shock wave is produced—a kind of sonic boom in the rock. The shock wave from such an impact crushes and compresses the impactor and target rock so intensely that after the shock passes, the decompressing rock will fly apart, or melt, or even vaporize. The concept of rocks instantaneously boiling away to vapor conveys a gut feeling for the extraordinary and violent conditions during an impact.

Scientists immediately ask about the energy of the approaching object, because energy is Nature's currency, a measure of the ability to move things around and bring about changes.[4] Nature runs a kind of automatic bookkeeping system for energy transfers, requiring that the energy of motion of the incoming comet be fully accounted for in all the kinds of damage done during the impact. When we do the bookkeeping, we find that the energy of motion of the comet just before impact was equivalent to the explosion of 100 million megatons of TNT, sufficient to vaporize the comet in about 1 second and to blow out a hole in the ground which was briefly 40 km deep but quickly collapsed into a broader, shallower crater 150–200 km across. To get a feeling for this quantity of energy, keep in mind that one large hydrogen bomb has a yield of about 1 megaton of TNT, and that the total nuclear arsenal of the world at the peak of the Cold War was about 10,000 such bombs. The 10^8-megaton impact of the comet which ended the Cretaceous was therefore equivalent to the explosion of 10,000 times the entire nuclear arsenal of the world (although the impact explosion was not nuclear).

Returning to the 10-km-wide comet as it approached the Earth at 30 km/sec, we can get a feeling for how fast the event happened. An airliner flies at an altitude of about 10 km, so imagine a plane unfortunate enough to be in the way of the incoming comet. In an instant the airplane would be smashed like a bug by the onrushing body. One-third of a second later the front of the comet, carrying the insignificant aircraft wreckage, would hit the ground, generating a blinding flash of light and initiating shock waves in the comet and the ground, and after another ⅓ second the back end would be passing below ground level. By one or two seconds after the loss of the airplane, there would be a huge, growing, incandescent hole in the ground and an expanding fireball of vaporized rock, and debris ejected by the explosion would be clearing the atmosphere on its way to points around the globe. Earth would suffer cataclysmic damage in less time than it takes to read this sentence.

Now that we have some sense of the scale of the impact that ended the Cretaceous world, let us look at our current, imperfect understanding of just what happened.

THE MOMENT OF IMPACT

The comet approaching Earth 65 million years ago first encountered the tenuous air many kilometers above the surface. About 95 percent of the atmosphere lies below an altitude of 30 km, so depending on the velocity and the angle at which the impactor approached the surface, it would have taken only a second or two to penetrate most of the atmosphere. The air in front of the comet, unable to get out of the way, was violently compressed, generating one of the most colossal sonic booms ever heard on this planet. Compression heated the air almost instantaneously until it reached a temperature 4 or 5 times that of the Sun, generating a searing flash of light during that one-second traverse of the atmosphere.

At the instant of contact with the Earth's surface, where the Yucatán Peninsula now lies, two shock waves were triggered. One shock wave plowed forward into the bedrock, passing through a three-kilometer-thick layer of limestone near the surface, and down into the granitic crust beneath. The onrushing shock wave drove forward through the bedrock, crushing shut all cracks and pore spaces and destroying much of the orderly crystal structure of minerals.

Meanwhile, a second shock wave flashed backward into the onrushing comet. Reflecting off the back of the impactor, it tore apart the trailing edge of the comet. In the second or so it took for this to happen, the comet ceased to be recognizable as a spherical body. With its enormous momentum driving it forward, the comet penetrated deep into the Yucatán bedrock, forcing open a huge hole and molding itself into an incandescent coating on the inside of the growing hole, which was now opening out into an expanding crater. But the comet coating on the inside of the crater did not last more than a moment before it was mostly vaporized, along with much of the original target rock.

As the rapidly vaporizing comet wreckage was carried forward into the growing crater, the shock wave curved back up to the surface and spewed out ejecta—melted blobs and solid fragments of target rock—upward and outward on high, arching trajectories that flung them through the thin outer fringes of the atmosphere and beyond. Falling back to Earth within a few hundred kilometers of the rim of the crater, this debris built up a vast blanket of ejecta.

Even this did not exhaust the pyrotechnic potential of the impacting comet. The huge cloud of vaporized rock generated at ground zero was driven outward by its own heat and pressure in a colossal fireball. The explosion of a nuclear bomb—tiny by comparison—produces a hot-gas fireball which flashes outward to a diameter of a kilometer or so, until it can push no farther against the atmospheric pressure, and then floats upward to an altitude of 10 km where it spreads out into a mushroom cloud. The incomparably greater fireball of the Yucatán impact overwhelmed the atmosphere, blowing right through the entire blanket of air, expanding and accelerating out into space and launching particles of rock into trajectories which carried them far around the Earth before they fell back to the ground.

And still the fireworks continued. Even as the scorching fireball of rock vapor blew away into outer space, it was followed by a second fireball, not as hot, but almost as dramatic. For about three kilometers down from the surface, the Yucatán was covered with a thick layer of limestone. Limestone is Nature's way of storing carbon dioxide gas as a solid, by combining it with calcium. Shocked limestone suddenly releases its stored CO_2, and in an impact as large as this, enormous quantities of this gas were almost instantaneously released like popping the cork on a colossal bottle of champagne. Still more rock debris was carried aloft in this second exploding gas ball as it, too, blew through the atmosphere and into outer space.

Meanwhile, the expanding crater had reached its maximum depth of perhaps 40 km. This hemispherical "transient cavity" was far too deep to be supported by the relatively weak rock of the Earth's crust, and the center began to rise, even as the perimeter continued to expand. While the steep outer walls collapsed in giant landslides, deep rocks from the mantle, below the granitic crust, rebounding after the passage of the shock wave, rose upward faster and faster into a central peak like those preserved in many craters on the Moon. The central peak of the Yucatán crater was so large and high that it in turn collapsed downward, driving outward into a set of ringlike ridges which left a pattern resembling a bull's-eye imprinted on the Earth to mark the site of this cataclysmic event.

THE RING OF DEVASTATION

In the zone where bedrock was melted or vaporized, no living thing could have survived. Even out to a few hundred kilometers from ground zero, the destruction of life must have been nearly total. Sterilized by the intense light from shock-compressed air and from the fireball of rock vapor, crushed when pores and cracks in rock were slammed shut by the passing shock wave, and bombarded by the falling debris of the ejecta blanket, little or nothing was left alive in this central area.

Out to a few thousand kilometers, into the area of modern Mexico and the United States, the Yucatán impact sent dramatic messengers of destruction. Animals living just over the horizon first witnessed a flash of light in the sky, then a last moment of calm. Then, as the ground began to shake uncontrollably from the passing seismic waves, the sky itself turned lethal. Beginning with a faint glow, the sky grew more and more intensely red, passing into incandescence, growing brighter and brighter, hotter and hotter. Soon the Earth's surface itself became an enormous broiler—cooking, charring, igniting, immolating all trees and all animals which were not sheltered under rocks or in holes. This fearsome phenomenon was produced by ballistic ejecta particles blasted into space by the impact, which were now falling back to Earth, reentering the atmosphere, heating up through friction with the air, and transmitting that heat to Earth as infrared light.[5] Only places which happened to be shielded by thick storm clouds would have avoided this lethal heat. Entire forests were ignited, and continent-sized wildfires swept across the lands. The ejecta particles had barely fallen to Earth and the lethal, incandescent sky returned to normal, when the air was blackened by rising plumes of soot from fires which were consuming the forests and removing the oxygen from the atmosphere.[6]

Even as the forests were set ablaze, another horror was approaching the coasts of the Gulf of Mexico. The impact occurred in the shallow water and coastal plains which flanked the Gulf, but it produced a huge disturbance in the waters of the deep Gulf, through seismic shaking, submarine landslides triggered by the seismic waves, and by the splashdown of the ejecta blanket. The result was a gigantic tsunami[7]—a massive wave perhaps a kilometer high, which spread outward across the Gulf of Mexico at terrific speed. Everyday waves do not disturb the bottom of deep seas like the Gulf, which are the quietest, calmest places on Earth. But the impact tsunami was so enormous that its keel swept across the bottom of the Gulf, digging channels into the fine sediments of the sea floor, and mixing them with the impact debris which had just fallen. As the tsunami front reached the shallow water of Florida and the Gulf Coast, it was pushed up higher and higher into a wall of water that towered above the shoreline. As this deluge crashed onto the coast, it not only ripped apart whole forests, but it shook the continental margin so violently that huge volumes of sediment were mobilized into submarine landslides which flowed down into the deep Gulf, burying the impact debris which had only just fallen.

Within hours of the impact, most of Mexico and the United States must have been reduced to a desolate wasteland of the most appalling, agonizing destruction. Where only the day before there had been fertile landscapes, full of animals and plants of all kinds, now there was a vast, smoldering netherworld, mercifully hidden from view by black clouds of roiling smoke.

Farther away from the Yucatán, the effects were less dramatic. The giant tsunami was largely confined to the enclosed Gulf of Mexico and could not reach Asia, Africa, or Europe. Ejecta particles rained down around the world, but fewer particles traveled to more remote areas, so the firestorms may not have been as intense as in North America. In contrast to the largely sterilized regions close to ground zero, distant continents may have escaped the direct effects of the Yucatán event. Tragedy would unfold more slowly in these remote areas, through the secondary effects of the impact.

THE HORSEMEN OF THE APOCALYPSE

Terrible as the immediate, direct effects of the impact were in the surrounding region, they probably would not by themselves have caused the disappearance forever of whole families of plants and animals, because survivors in remote regions would have repopulated the devastated regions in the years to come. And yet an enormous mass extinction did follow the impact, and we now understand some of the longer-term global disasters which were secondary results of the impact. Let us review these Horsemen of the Apocalypse in their order of appearance.

Within days of the impact, the immediate effects had died down. The fires were probably going out, the tsunami had spent its main strength against the coast of the Gulf of Mexico, and violent winds were settling down. But the Earth was turning cold and dark. Vast quantities of fine dust had burst through the atmosphere in the fireball and the dust was now settling through the upper atmosphere around the world, blocking the sunlight. The land became so dark that you could not have seen your hand in front of your face, and this darkness and the accompanying cold probably lasted for a few months, until finally most of the dust had settled to the ground.[8]

But after the return of light, the climate went to the opposite extreme. Two greenhouse gases—water vapor and carbon dioxide—had been released in vast quantities from the site of the impact. The water vapor was probably removed quickly from the atmosphere as rain which washed out the dust. Carbon dioxide can only be removed slowly from the air, and now it trapped the heat from the Sun, raising temperatures to sweltering levels. It was probably thousands of years before the carbon dioxide was back to normal levels.

Not only were water and dust raining out of the atmosphere, but there was also a devastating acid rain.[9] Some of this may have been sulfuric acid derived from sulfur in anhydrite, a sedimentary rock interbedded with the limestones of the Yucatán. But much was nitric acid, originating from the atmosphere itself. The air we breathe is about 20 percent oxygen and most of the rest is nitrogen. Normally these occur as two-atom molecules of oxygen, O_2, and of nitrogen, N_2. Nitrogen forms very stable molecules which are tightly bonded together. Only when the air is strongly heated are the N_2 molecules broken up, allowing some of the nitrogen to combine with oxygen as molecules of nitric oxide, NO. This happened on a grand scale during the impact event when the air was heated by shock waves, by the fireball, and by the friction of reentering ejecta. Vast quantities of nitrous oxide were formed, which reacted with oxygen and water vapor in the atmosphere to form nitric acid, HNO_3, which rained out of the sky, killing plants and animals and dissolving rocks.

A world first dark and frozen, then deadly hot, a world poisoned by acid and soot. This was the global aftermath of the Yucatán impact. We wonder how anything could survive this environmental apocalypse. Yet there were survivors, and their descendants populate the world today.

VICTIMS, SURVIVORS, AND DESCENDANTS

By the time the physical devastation caused by the impact had faded, years or centuries after the event, Earth's biosphere was changed forever. Whole groups of plants and animals had disappeared, never to be seen again. By one estimate, half of the genera living at the time of the impact perished. This was one of five great biological mass extinctions we know of in Earth's past. It is very difficult to learn what caused the loss of any particular group of plants or animals. Some reasonable inferences have been made, but in many cases we will probably never know with certainty. It is easier to construct the list of victims and survivors.

Best known of the victims, of course, are the dinosaurs. *T. rex* and the other huge carnivores perished, as did the herbivorous dinosaurs, as well as their relatives who swam, like the mosasaurs, or flew, like the pterodactyls. Most paleontologists now consider that modern birds are very closely related to the dinosaurs which, in this sense, did survive the end of the Cretaceous.[10] Yet recently discovered fossils are revealing that birds were nearly wiped out as well.[11]

The loss of the dinosaurs is probably related to their position in the food chain, with herbivorous dinosaurs eating vegetation and carnivorous dinosaurs eating herbivores and perhaps small mammals. During the months of cold and darkness cast by the pall of dust in the atmosphere, plants would wither and the herbivores would starve, and so would the carnivores in their turn. Large animals are never abundant, especially top carnivores, so they would have been particularly vulnerable to extinction.

Many smaller land animals survived, including mammals, as well as reptiles such as crocodiles and turtles. No one really understands why these animals escaped extinction. Being smaller and thus more numerous would increase their chances of survival, and this may help explain the survival of birds as well.

Leaf fossils demonstrate that land plants also suffered a mass extinction.[12] We expect that individual trees and bushes alive at the time of the impact would have perished in the cold and the dark. But seeds and roots should have allowed most species to reappear after the darkness ended. The extinction of many kinds of plants has not been explained.

Turning to the less familiar marine realm, we find that the impact spelled the end of the coiled-shell ammonites—relatives of the chambered nautilus—which had flourished in the seas for hundreds of millions of years.[13] Lesser known groups of invertebrates perished wholesale at the level of genera and families. Perhaps they were the victims of food-chain collapse, or perhaps their shells were dissolved in acidified seawater, but no one knows.

Still less familiar are the microscopic single-celled plants and animals that float in the surface waters of the ocean. These tiny organisms were enormously abundant but suffered nearly complete extinction. The microscopic photosynthetic algae and the single-celled predators called foraminifera produced vast numbers of tiny platelets and miniature shells that record the mass extinction with unusual clarity.[14] Probably vulnerable to darkness and acid, they were the base of the marine food chain, and their loss was devastating to marine animals that depended on them. Both foraminifera and photosynthetic algae were at grave peril, with many or most species perishing, but in both cases a few species survived and left descendants which abound in the oceans today.

The sudden loss of half the genera of plants and animals on Earth is a catastrophe almost incomprehensible to us. It truly marked the end of a world. And yet, the darkness eventually faded, the heat died down, and the acids were neutralized. Survivors there were, and they found themselves in a new world, tragically changed, but with boundless opportunities for the future. For 150 million years dinosaurs had been the large land animals of the planet while mammals were confined to the role of small animals. With the disappearance of the dinosaurs, there were new opportunities for mammals, and evolution rapidly produced large ones. Our nostalgia for the lost world of the Cretaceous is tempered when we realize that it was a world that held no place for us—for large mammals. Our horror at the destruction caused by the impact that ended the Cretaceous is eased by the understanding that only because of this catastrophe did evolution embark on a course which, 65 million years later, has led to us. We are the beneficiaries of Armageddon.

JUST HOW DO WE KNOW ALL THIS?

Tolkien's story of Middle Earth is, of course, pure fantasy. It has its own internal logic, but magical things take place in Middle Earth which could never happen in the real world. It is a wonderful story, but in order to enjoy it, you must suspend your sense of disbelief. It is not in any way intended to recount events which ever really occurred.

The story of the impact on the Yucatán which ended the age of the dinosaurs has a different purpose. It is intended to be a reconstruction, as accurate as possible, of historical events which really did happen. It asks not that its readers suspend disbelief, but that they do exactly the opposite—that they bring to it their most critical facilities, that they search it for flaws, that they test it in any way they can and try to improve its accuracy.

But how can we possibly reconstruct events which happened 65 million years ago, long before any human being was around to observe what happened and record it for posterity? We can reconstruct these events because the history of the Earth is recorded in the Earth itself. Most of the history of our planet is written in rocks. Rocks are the key to Earth history, because solids remember but liquids and gases forget. Retrieving these long-lost memories is the business of geologists and paleontologists, of people who have chosen to be the historians of the Earth.

Understanding how we decipher a great historical event written in the book of rocks may be as interesting as the event itself. Uncovering the extinction that ended the Cretaceous has been a saga of patient detective work, of high adventure in remote parts of the world, of lonely intellectual struggle, of long periods of frustration ended by sudden breakthroughs, of friendships made or lost, of the embarrassment of public mistakes and retractions, of the exhilaration of discovery, and of delight in a wonderful emerging story. This is what we will explore in the rest of this book, as we see how the story of the Yucatán impact was uncovered and pieced together.

CHAPTER 2

Ex Libro Lapidum Historia Mundi

HISTORY WRITTEN IN ROCKS

As recently as 1975, the story of the impact on the Yucatán was completely unknown. One of the most dramatic episodes in the past of our planet had been absolutely forgotten, lost beyond memory for 65 million years. How has this lost memory been recovered?

We are born in ignorance of the events that took place before our birth, and through the study of history we seek to overcome this native amnesia. We can hear about the most recent events by asking our parents and grandparents, who remember them. History from the times before living memory is written in documents, both the original writings of people long gone and the books written by scholars of history. Through the words represented by symbols on paper we are carried back through 5,000 years, back to the earliest writings, learning the thoughts and deeds of the people who lived before us.

Yet 5,000 years takes us back only a *millionth* part of the lifetime of the Earth. Back beyond the invention of writing stretches an almost endless abyss of time, during which the events took place which determined the kind of creatures we are and the kind of world we live in. It is only in the last couple of centuries that we have learned to decipher the events of this forgotten eternity and to write down its history.

The key discovery was that history is written in rocks. *Ex libro lapidum historia mundi*—from the book of rocks comes the history of the Earth. More people are fascinated with plants and animals than are drawn to rocks, because plants and animals are living and dynamic, while rocks seem to lie there inert and unchanging. Rocks do change, but usually so slowly that few people notice the changes. It is precisely this sluggish, nearly static character that makes them good recorders of Earth history. Rocks remember the past.

Recovering the memories held in rocks is most easily understood in the work of archaeologists. Ancient temples, buildings, and cities record the lives of civilizations long gone, like that of the ancient Mayas, which flourished, declined, and disappeared in the jungles of the Yucatán, above a buried crater dating from a much earlier extinction.

Archaeological sites the world over display the fundamental rule of history written in rocks—younger layers rest on older ones. This is the law of superposition, the basis of all stratigraphy. Stratified rocks are deposited in succession, layer upon layer. Of course one must be careful of exceptions where pits have been dug down into older deposits and filled with younger material, or where caves have been hollowed out and then filled up, but these occasional possibilities for making mistakes just help keep the stratigrapher alert!

Archaeological sites that have been occupied for centuries are rich in examples of superposition. In Rome, for example, the Forum preserves layer after layer of history, with pre-occupation sediments of the Tiber River overlain by the primitive remains of archaic Rome, then by ruins of the stouter buildings of Republican Rome and the majestic monuments of the Roman Empire. Medieval and modern Rome has been built over multiple layers of ruins, and sometimes there are steps leading down from today's street level to the entrance of an ancient church. Stratigraphy underlies every ancient city.

History is written in rocks at a more detailed level in stone buildings everywhere. North of Rome, in the Apennine Mountains, the beautiful little city of Gubbio quietly dreams of centuries past in a setting of noble medieval stone towers, churches, and palaces. I have spent many months doing geological research in the mountains near Gubbio, and I never tire of walking through its streets, looking at the medieval buildings. The entire city is a history book written in stone, and looking closely,

Medieval architecture in Gubbio—the Palazzo dei Consoli.

one finds historical episodes captured and preserved in the details of construction and reconstruction. For instance, pointed Gothic windows are everywhere in Gubbio, but often they were walled up, and later reopened with windows of different shapes in different places. I have been told that this sad defacing of noble architecture records the imposition of a tax on windows. Some owners of houses chose to avoid the tax by walling up the windows, and by the time the tax was finally rescinded, the original locations had been forgotten, or the interiors had been changed, or Gothic windows were no longer in style. To anyone fascinated with the Middle Ages, and to anyone with an eye for history written in stone, Gubbio is a paradise.

READING EARTH HISTORY

Human history written in stone walls and in the layers of archaeological ruins is easily grasped by most of us, because we are used to buildings and may have built them ourselves. The rock record of prehuman events is not so familiar; it takes study and experience to learn to read Earth history. But Gubbio is a good place to start. Walking out the postern gate in the town wall, into the mountains behind the city, we can see how Earth history is recorded.

Outside the gate we pass a few outlying stone houses as the rocky mountain sides converge, closing us into a canyon called the Gola del Bottaccione. "Bottaccione" means "big water barrel" in Italian, and it is a whimsical name for a medieval aqueduct built in the fourteenth century to bring water from a mountain spring down the canyon to Gubbio. The aqueduct snakes along the mountainside above the modern road, and in the road cuts there are outcrops of an attractive pink stone—the Scaglia rossa. ("Scaglia" is pronounced "Scáhl-yah" in Italian. It means scale or flake and refers to the ease with which this rock can be chipped into handsome building stones. "Rossa" refers to the red color.)

The Bottaccione Gorge at Gubbio. The massive mountainside in the middle distance is the Cretaceous part of the Scaglia rossa limestone. The horizontal structure above the road is the medieval aqueduct.

Pausing to look carefully at the Scaglia rossa, we first notice that it is arrayed in layers or beds about 10 cm thick, which dip, or slant, about 45° away from Gubbio. The Scaglia is a sedimentary rock formed by the deposition of particles of sediment on the sea floor and later was pushed up to form the Italian Peninsula. In this case the particles are mostly grains of the mineral calcite (calcium carbonate, or $CaCO_3$), forming the sedimentary rock called limestone. Sedimentary rocks are deposited in roughly horizontal layers, so the 45° dip shows that these beds have been tilted after their deposition. This tilting is due to the episode of deformation which produced the folds of the Apennine Mountains. Here is a piece of history written in rocks, and it takes us beyond archaeology, for buildings may collapse but they are rarely folded!

These originally horizontal beds of pink Scaglia limestone continue for a thickness of 400 meters, and there are more limestones of other colors above and below. Clearly there is much Earth history recorded in these layers, but what kind of history? To find out, we break off a fresh piece of Scaglia and look at it with the little magnifying hand lens all geologists carry. Tiny specks throughout the rock now resolve themselves into little coiled, chambered microfossils. These are the shells of foraminifera, the single-celled predators that float near the surface of the deep oceans, and which almost perished, although not quite, in the same mass extinction that finished off *T. rex*. The presence of foraminifera shows the Scaglia must be a marine limestone. We can also tell that it is a deep-water limestone, because there are none of the invertebrate fossils that abounded in the shallow seas of that time. And it was deposited far from the mouth of any river, because it is almost free of sand and silt.

Perhaps many people would find this submarine limestone less interesting than a sediment deposited on dry land—after all, we are land animals, and dry land and its inhabitants are more familiar and generally more significant to us. But land above sea level is the main site of erosion, which levels hills and mountains and removes sediment previously deposited. Erosion destroys the record of Earth history. On the deep sea floor, however, there is little erosion because waves cannot reach the bottom and currents are slow and gentle. Deep-sea sediments are ideal recorders of Earth history, and the limestone at Gubbio is one of the very best historical sequences in all the world.

Since this was recognized in the mid 1970s, geologists like me have come to Gubbio each summer, seeking the answers to many different questions about Earth history. It was at Gubbio that the first hint of the great impact came to light, and we will turn to that soon. But first we need a deeper understanding of the vast stretches of time that lie far back before the appearance of human beings.

THE MEASURES OF EARTH HISTORY

Scholars of human history have two ways of identifying times past. Sometimes they give numerical dates: "Rome was founded in the 8th century B.C., and its last recognized emperor was deposed in A.D. 476." Sometimes they refer to named periods of human history: "Gubbio flourished in the High Middle Ages, and its architecture represents the Italian Gothic period."

Geologists also use both of these systems: "The Yucatán impact 65 million years ago marked the boundary between the Cretaceous and Tertiary periods." It takes some effort to get used to the two ways of specifying times in the Earth's past, because the names are unfamiliar and the numerical dates in millions of years seem unimaginably remote.

When geologists 200 years ago first began to understand Earth history as recorded in rocks, they gave names to recognizable historical intervals based on the character of the rocks. Many of the names have stuck because they are useful. Just as an art historian can recognize a Gothic window even though the exact date of the building may be unclear, so a geologist can often tell from fossils of ammonites or foraminifera that a rock is of Cretaceous age even though there is no way to determine its precise age in years.

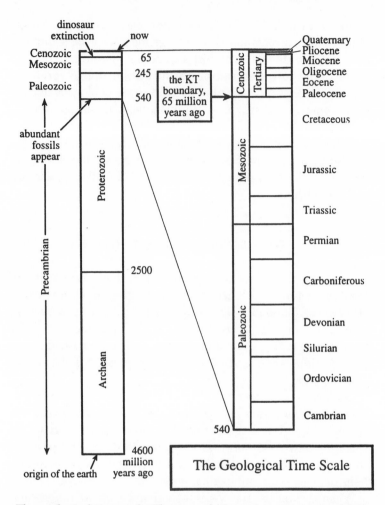

The Geological Time Scale

The geological time scale. These are the most important named intervals in Earth history. The column at the left, with older at the bottom, shows the entire history of the Earth. The column at the right shows the more detailed subdivisions of the last 12% of Earth history, made possible by abundant fossils.

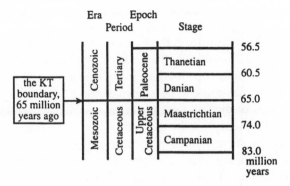

These names are used by geologists for time intervals in the vicinity of the boundary between the Cretaceous and Tertiary periods, which is usually called the KT boundary. The sequence from left to right is from longer time intervals to briefer ones, and older time intervals are at the bottom.

THE KT BOUNDARY

The names of the geologic time intervals are given in the table on page 26, but since we are dealing in this book only with the great extinction event 65 million years ago, we need remember only the intervals which it separates, as summarized in the table on this page.

In terms of eras—the broadest divisions of Earth history—the Yucatán impact separates the Mesozoic ("middle life," or the Age of Dinosaurs) from the Cenozoic ("recent life," or the Age of Mammals).

In the more detailed subdivisions called periods, the impact event marks the boundary between the Cretaceous and the Tertiary periods. The name Cretaceous comes from the Latin "creta," meaning chalk, because chalks were deposited in shallow seas over broad areas during the last third of the Mesozoic. The name Tertiary is left over from an early idea that there had been four intervals in Earth history—the primary, secondary, tertiary, and quaternary ages. The first two names have disappeared, but Tertiary has been retained for the time between the end of the Cretaceous and the beginning of the Ice Age, which is called the Quaternary. Geologists use the letter K to symbolize the Cretaceous, from the equivalent German word "Kreide" (chalk), and T to indicate the Tertiary. The mass extinction at the end of the Cretaceous marks a major turning point in Earth history which has come to be known as the KT boundary.

In terms of the still finer historical subdivisions called stages, the KT boundary marks the end of the Maastrichtian, the last Cretaceous stage, named for rock exposures around the Dutch city of Maastricht, and the beginning of the Danian, or first stage of the Tertiary, named for Denmark where rocks of this stage are well exposed.

UNDERSTANDING DEEP TIME

Learning the intervals of Earth history is simply a matter of making the effort to memorize some unfamiliar names. Appreciating the numerical dates is more difficult, because intervals of millions of years are hopelessly beyond the comprehension of human beings who may, with luck, live for a single century. A geologist has no more intuitive, gut understanding of these immense stretches of time than does anyone else. But geologists speak of millions of years with familiarity because we know what happened when in Earth history, and because we know which dates are geologically recent and which are geologically ancient. To achieve this familiarity we need to do two things—we need to change our units of time from "years" to "million-years," and we need to recall the ages in million-years of the fundamental divisions of Earth history.

In thinking about Earth history, we need to change our units of time from years to "million-years," because thinking of Earth history in terms of years is as hopeless as measuring a trip from Mexico to Italy in centimeters, or measuring a lifetime in seconds. So we need to think of the date of the dinosaur extinction not as an enormous number of years—"65 *million* years!"—but as a small number of "million-years"—just 65 of them.

Then we need a way to appreciate how far back in Earth history that date lies. Here we are helped by a remarkable coincidence. Earth history is, by chance, about one million times as long as the written record of *human* history. Writing was invented about 5,000 years ago, and the Earth was formed about 5,000 million-years ago.[1] So we can recognize the recentness or antiquity of a date in Earth history given in million years by comparing it to a date in human history in years. The Quaternary ice ages began 2 million years ago, which in the sweep of Earth history is as recent as a human event just 2 years ago. The Cretaceous-Tertiary boundary, 65 million years ago, is the Earth equivalent to a human event 65 years ago, within the memory of many people. Just as human history seems really remote when we get to events many hundreds of years ago, so Earth events many hundreds of million years ago belong to the really distant past.

These are the ways of thinking that young geologists absorb when they begin to study Earth history—they learn the names, change their units from years to million years, and come to appreciate which dates are relatively recent and which are really remote.

THE CHRONOLOGY OF EARTH HISTORY

How can we assign particular rocks to named periods of Earth history and learn how old they are in millionyears? The *sequence* of ages is usually easy to determine, using the law of superposition. The named periods came early in the development of geology as a science and were based on fossils found in the rocks, as they still are today, two hundred years later. Determining numerical ages in millions of years is a twentieth-century achievement which was not possible until physicists had discovered radioactivity and invented sophisticated analytical instruments.

There still are major limitations in our ability to work out a full chronology of Earth history, because numerical ages and fossil ages usually come from very different kinds of rock.[2] Most of the radioactive minerals that yield numerical ages form at high temperature during the crystallization of igneous rocks from molten magma, whereas fossils are found in sedimentary rocks deposited in the ocean or on land, at temperatures conducive to life.

Geologists have only slowly made progress in bridging this gap and building a time scale with accurate numerical ages attached to the boundaries between the named intervals based on fossils. The most direct approach has been to find datable high-temperature minerals in volcanic ash that has blown far from the erupting volcano and settled as layers in sedimentary rocks that contain fossils.[3] An indirect but very effective approach has been to develop a third time scale, based on reversals of the Earth's magnetic field, which are recorded in both igneous and sedimentary rocks. Every few years a book is published presenting the current state of knowledge of geochronology—of the dating of Earth history.[4] The rock sequence at Gubbio has been important in all three major approaches to geochronology, and time-scale work at Gubbio led directly to recognition of the impact at the KT boundary, so let us look more closely at those three methods of dating rocks as applied to the Gubbio rock sequence.

DATING ROCKS WITH FOSSILS

People must have found fossils long before it was generally accepted that they are the remains of animals and plants which lived long ago. William Smith was an English canal engineer who spent the years around 1800 digging ditches in sedimentary rocks. Smith gradually became familiar with the kinds of fossils his workmen would find, and he recognized that fossils change in a recognizable way through the long stretches of time recorded in sedimentary rocks. He realized that fossils can be used to place sedimentary rocks in a chronological sequence and discovered that it was possible to correlate rocks of the same age over long distances. This has been the basis of stratigraphic paleontology ever since.

It is an observational fact that animals and plants preserved as fossils in sedimentary rocks change as we pass upward through the rock sequence. Evolution in this sense can be confirmed by anyone who looks carefully for and at fossils. Through the nineteenth century, names like Cretaceous and Tertiary were applied to rocks of particular age ranges as the sequence of changing fossils was gradually worked out. The reason for the changes in fossils was mysterious until English naturalists Alfred Russel Wallace and Charles Darwin explained them as the result of natural selection. As we shall see, Darwin insisted that all evolutionary change has been gradual. The theory of evolution by natural selection has held up well for over a century, but some details of the theory may still need improving.[5] One focus of this book is how Darwin's insistence on gradual evolutionary change has been challenged through recognition of the dramatic effects on life of occasional catastrophic events like the impact on the Yucatán at the time of the mass extinction at the KT boundary.

Fossilized marine invertebrates like clams, ammonites, and corals were the most useful dating tools in the nineteenth century because they were easy to find in the field and big enough to study with the unaided eye. In the twentieth century, paleontologists came to appreciate the usefulness of the tiny "microfossils" which occur in profusion and can be recovered in large numbers, even from drill cores which are unlikely to intersect the rarer large fossils. Although a microscope was required for studying them, microfossils became the dating material of choice by the middle of the twentieth century.

The most important of the microfossils are the foraminifera, or "forams" for short. These single-celled marine organisms make tiny shells which are different for each species and can be identified accurately under the microscope. Most forams live on the sea bottom, and the species present in a particular sedimentary rock thus reflect primarily the sea-floor environment. But some forams float as plankton in the surface waters of the ocean.

These planktic forams are particularly useful for dating rocks, because ocean currents spread newly evolved species rapidly throughout the oceans of the world, and thus evolutionary changes are immediately recorded worldwide when the forams die and their shells settle to the bottom. Only in extremely deep-water marine sediments are foram shells absent, because they dissolve in the very cold water of the deep ocean.

It was not until the 1960s that paleontologists fully realized the value of the nearly continuous historical record preserved in limestones which accumulate on the ocean floor at middle depths. Called "pelagic" limestones to distinguish them from "neritic" limestones which are built up by the fossils of organisms that live on the sunlit bottom in shallow water, these sediments are deposited in the darkness far below the deepest eroding waves and lie undisturbed for tens of millions of years. Many pelagic limestones of Cretaceous and Tertiary age are packed with planktic forams, so they can be dated in detail.

Beginning in 1967 the deep-sea cores recovered by the scientific drilling ship *Glomar Challenger* yielded a cornucopia of information on Earth history. Not all deep-sea limestones are still submerged in the ocean, however—in a few places they have been pushed up above sea level and exposed in the mountains, and can be studied by those who can't afford a drilling ship but do have boots, hammer, and hand lens. But it took a while for geologists even to be sure that these limestone exposures were of deep-water, pelagic origin, let alone to appreciate their value as records of Earth history.

Extensive outcrops of pelagic limestones are rare. One of the few good places to find them is in the Apennine mountain range of Italy, and perhaps the best place in the Apennines is the Bottaccione Gorge at Gubbio. As a student in Milan in the 1960s, Isabella Premoli Silva studied the forams of the Scaglia rossa at Gubbio,[6] learning to identify them even though the Scaglia is a hard limestone, and you cannot get the forams out intact. Many paleontologists are good at identifying loose forams, but Isabella is one of the few who can identify them in thin sections—in slices of rock mounted on glass slides and ground down until the rock is transparent. Most of us can identify our friends' heads when we see them, but it is much harder to identify a head in silhouette, as Isabella can do with forams. With her ability to recognize forams in hard limestone and with the continuous 50 million-year record of the Scaglia rossa, Isabella was able to recognize in the Gubbio limestones the sequence of stratigraphic age zones established elsewhere on the basis of forams extracted from soft sediments.

NUMERICAL AGES IN MILLION YEARS

To go beyond the paleontological time scale and names based on the occurrence of fossils, it is critical to determine the numerical ages of events in Earth history. How long ago did the KT boundary extinction take place, in million years?

The determination of numerical ages is based on the decay of radioactive atoms. A few of the elements that occur in minerals have unstable nuclei which change into other elements. This offers us a clock, because as time passes and the atoms of the parent element in a mineral grain gradually change into the daughter element, the ratio of daughter to parent increases. Radioactive decay takes place at an unvarying rate, no matter what changes in pressure or temperature, or what chemical reactions go on, so it provides a very reliable clock. Uranium, thorium, rubidium, and samarium are all unstable and gradually decay, but the most important for dating sedimentary sequences is potassium, which decays to argon. The decay rate for radioactive potassium has been determined, and if one can measure the amount of the parent potassium and the amount of daughter argon, it is possible to calculate the age of a sample in million years. The greater the ratio of daughter argon to parent potassium, the older the rock.

In practice this radiometric dating is a complicated and sophisticated task, with all kinds of possibilities for error. For example, argon is a gas and sometimes the daughter argon leaks out of a sample over millions of years, making it appear too young. Or if some argon was incorporated in the mineral when it first formed, the apparent date will be too old. Yet the geochronologists, or age-daters, who do this work have become very sophisticated and skilled, and many accurate age dates are now published each year.

Our present understanding of the time scale of Earth history comes partly from fossils and partly from numerical ages tied to the fossil information. But another important key to developing the time scale has been the study of reversals of the Earth's magnetic field. This is how Bill Lowrie and I got involved in time-scale research, and how we were introduced to the mysterious extinction of *T. rex*.

FOSSIL COMPASSES AND THE REVERSING MAGNETIC FIELD

Bill Lowrie is a geophysicist from the southern Scottish town of Hawick. He and I came to Lamont-Doherty Geological Observatory, Columbia University's oceanographic and geological laboratory, at about the same time, in the early 1970s. As hungry young researchers we were looking for exciting projects, and we started sharing ideas.

I told Bill that I was interested in deciphering the origin of the Apennine Mountains in Italy, where I had been working, and he told me what could be done with paleomagnetism, his specialty in geophysics. Some rocks contain magnetic mineral grains which record the direction of the Earth's magnetic field at the time they were deposited as sediments, or cooled as lava flows. These magnetic mineral grains act like hidden fossil compasses, and in the laboratory paleomagnetists can read those fossil compasses.

Paleomagnetism had been critical in the plate tectonics revolution in the 1960s. Plate tectonics was a modern version of the idea of continental drift, and it swept away the old view that all continents had always remained in fixed positions. If continents had never moved, as most geologists believed in the first half of the twentieth century, then all fossil compasses in all rocks should still point north. The pioneering paleomagnetists had shown that this was absolutely not true. The fossil compasses—the remanent magnetization—often pointed in other directions because of rotation of the moving continents since the rocks were formed.

At the time Bill and I began our research careers, the theory of plate tectonics had just been accepted, and the rotations of large continental masses were at least roughly known from the early paleomagnetic studies. Having been an oil company geologist in Libya and a researcher in Italy for a few years, I had become interested in the complex geological evolution—the tectonics—of the Mediterranean, which seemed to involve motions and rotations of plates much smaller than the major continental masses. Bill and I started to talk about Mediterranean tectonics; we found ourselves using the term "microplates" and wondering what paleomagnetism might tell us about their motions. We realized that if the continental crust of Italy had rotated as a microplate during the deformation of the Apennines, we should find the fossil compasses in Apennine sediments rotated out of alignment, no longer pointing north.

So we made a trip to collect rocks in the Apennines, with our wives Milly and Marcia as field companions. In sunshine and rain, we criss-crossed the mountains, sampling the Scaglia rossa, because its rust-red color indicated the presence of the iron-oxide mineral hematite, which might record the Earth's magnetic field. Part of the time we worked with my friend Ernesto Centamore, a giant Italian with a gargantuan appetite for life, for food, and for geology. Ernesto took us to Isabella Premoli Silva's outcrops at Gubbio, which he claimed were the best exposures of Scaglia rossa anywhere. Indeed the Gubbio outcrops were spectacular, and we collected many samples up through the beds of limestone. We were hoping to see a progressive rotation during the time of deposition of the Scaglia, with the older fossil compasses twisted farther away from north than the

Bill and Marcia Lowrie near Gubbio, drilling samples for a paleomagnetic study.

younger ones. That would be the paleomagnetic signature of a rotation of the Italian microplate.

It had seemed like a fine idea, but unfortunately, when Bill measured the magnetization of the samples in the lab at Lamont, we realized that what we could learn was very limited. The fossil compasses pointed around to the west of north, showing that the crust of Italy had indeed rotated. But we couldn't work out the detailed history of rotation because the presence of bedding planes separating limestone beds had allowed the beds to twist around relative to one another during the Apennine folding that had given the limestone layers their 45° dip. The detailed pattern of remanent magnetization did not reflect the rotation of the Italian microplate; it was due to insignificant local disruption of the limestone.

This was a big disappointment to us. It seemed that we had wasted our time collecting all those samples. But then we discovered something that turned out to be much more important than measuring microplate rotations could ever have been. Although most of the Gubbio limestones had fossil compasses that pointed generally northward, a few of them pointed in exactly the opposite direction! Magnetic reversals had been discovered around 1960, and were a hot topic at Lamont, and Bill and I realized almost immediately that we were seeing a new kind of record of the reversals of the Earth's magnetic field. We were seeing a global phenomenon, something even more interesting than a local effect of the complicated tectonics of the Mediterranean.

It had been a real surprise to geophysicists, just a decade earlier, to find that at many times in the past, the Earth's magnetic field had reversed. The Earth acts as if it had a huge bar magnet in its interior, aligned roughly north-south, producing the global magnetic field which aligns the compasses on ships and the fossil compasses in rocks. But there is no bar magnet down there, for iron could not stay magnetized at the high temperatures deep in the Earth. The field is actually produced by swirling convective motions in the liquid iron core, which behaves like a magnetic dynamo. The early paleomagnetists found fossil compasses in young volcanic rocks that point north in some lava flows and south in others. After a long debate and many measurements they proved that the Earth's magnetic field has switched from pointing north to pointing south, and back and forth, over and over again for reasons that are still not understood. The rotation and the orientation of the Earth have not changed. Only the direction of its magnetic field has reversed.

Mysterious though they remain, magnetic reversals had been critical in testing the theory of plate tectonics a few years before Bill and I went to Gubbio. Proponents of plate tectonics had claimed that ocean basins grow by sea-floor spreading, with new ocean crust forming where submarine lavas cool at the crests of submerged mid-ocean ridges. And then it was recog-

nized that the ocean basins are imprinted with a magnetic striping, formed as cooling lava at the ridge crest traps a record of
the magnetic field direction, normal and reversed. The sea-floor
magnetic stripes were mapped by towing magnetometers behind ships and airplanes criss-crossing the world's oceans. The
ages of the last few magnetic reversals were dated in Hawaiian
lavas and corresponded to the widths of the youngest magnetic
stripes on the sea floor. It was the most important of the many
proofs of plate tectonics.[7]

The magnetic stripes on the sea floor continued on back into
older and older ocean crust. But there had been no way to date
the older reversals, and therefore no way to date the older ocean
crust. Those dates were needed in order to work out the history
of plate motions and continental drift. It had been a frustration
to the pioneers of plate tectonics, but as soon as Bill and I saw
those first reversed directions from Gubbio, we knew that the
key to dating the reversals was staring us in the face. After all,
the Scaglia rossa is full of forams—the best tool for dating marine sedimentary rocks, in the sense of placing them in the sequence of named historical intervals. We had just discovered
that the Scaglia also records magnetic reversals. In a round-
about, unexpected, very lucky way, we had gotten the break
that all young researchers dream of!

We made a new trip to the Apennines to take closely spaced
samples up and down the Scaglia rossa, to determine the detailed history of magnetic reversals. Almost immediately we
found that we were not the only ones who had realized that the
Scaglia rossa contained a record of magnetic polarity history. In
the alumni newsletter from Princeton I read that my former professor, Al Fischer, was doing the same thing, with two of his
current graduate students—Mike Arthur and Bill Roggenthen—
together with Isabella Premoli Silva and paleomagnetist Gio-
vanni Napoleone from Florence.

At first we were deeply disappointed. In science, simultane-
ous discoveries like this may lead to intense competition, which
is often beneficial, or to feuds, which are always malignant. To
avoid an argument over who made the discovery first, we got

Al Fischer describes the KT boundary outcrop at Gubbio to a group of geologists on a field trip.

together with Al's team and agreed to join forces and work together. In 1976, Paolo Pialli organized a conference to discuss the results, and in 1977 we published a set of five papers showing that the Gubbio Scaglia rossa records the same sequence of long and short polarity zones that is seen in the oceanic magnetic stripes.[8] The Earth has two tape recorders recording the polarity of the magnetic field, and they give the same reversal history, even though the sea-floor recorder runs 6,000 times faster than the deep-sea sediment recorder.

Over the next few years Bill and I, together with Al's group and a number of other colleagues, worked our way down through the Cretaceous and up through the Tertiary, determining the sequence of magnetic reversals and dating them with forams. Other paleomagnetists tried to do the same thing on pelagic sediments in deep-sea drill cores recovered by the *Glomar Challenger*, but they never seemed to be able to get a good reversal record, apparently because the vibrations of the drill bit loosened the soft pelagic mud of the cores, allowing the magnetic minerals to reorient. Later on, when quiet coring techniques were developed, our reversal sequence was confirmed from deep-sea cores, but for a few years in the mid-1970s it was only the hard pelagic limestones exposed on land which could be used to date magnetic reversals. Those of us who were working in the Apennines took advantage of that little window of opportunity, and after several detailed studies, Bill and I summed up the results in a paper entitled "One hundred million years of geomagnetic polarity history."[9]

SUBVERSIVE HINTS FROM A BED OF CLAY AT GUBBIO

Bill Lowrie and I returned to Gubbio several times in the mid-1970s, collecting more and more samples for determining the positions of magnetic reversals in the Scaglia limestones, in order to date them with Isabella's foram ages. Sometimes Isabella would come to work with us, and she showed us how to recognize the boundary between the Cretaceous and the Tertiary, which she had identified years earlier as a student. With a hand lens you could spot the near extinction of the forams, which are abundant and as big as sand grains in the top beds of the Cretaceous, but with only the very smallest ones surviving into the first beds of the Tertiary.

Bill and I learned to identify the KT boundary ourselves, and as we located this key break in outcrop after outcrop across the Apennines, we began to wonder about its significance. Why had the forams almost become extinct? What had happened to cause that extinction? And why was it so abrupt? At Gubbio, and in each new outcrop we found, there was a layer of clay about a centimeter thick, lacking fossils, between the last limestone bed with Cretaceous forams and the first limestone bed that

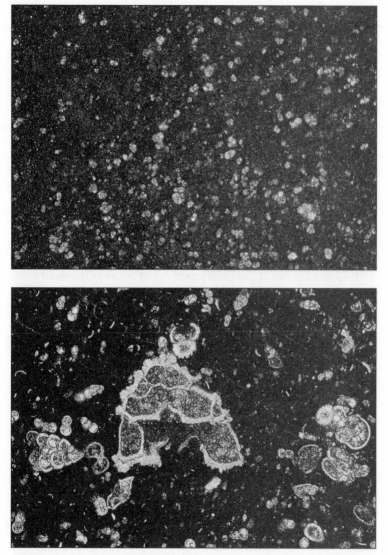

The near extinction of the planktic foraminifera at the KT boundary at Gubbio. The lower microscope photograph shows the large forams, up to 1 millimeter across, in the top bed of the Cretaceous. The upper photograph, at the same scale, shows the much smaller forams in the first bed of the Tertiary.

contained only the new Tertiary ones.[10] Did the clay have something to do with the extinction?

We invited Al Fischer to come to Lamont to give a lecture, and he stressed that the extinction of the marine microfossils that marked the KT boundary in the Gubbio limestones was at least approximately the same age as the most famous of all extinctions—the disappearance of the dinosaurs. The more I thought about the KT boundary, the more it fascinated me.

I remember very clearly walking around the grounds at Lamont one day shortly after Al Fischer's talk there and realizing fully that this was a world-class scientific problem. Much of the work we do as scientists involves filling in the details about matters that are basically understood already, or applying standard techniques to new specific cases. But occasionally there is a question that offers an opportunity for a really major discovery. Choosing what problems and what kind of problems to work on is a critical strategic decision for a scientist. The question of the KT extinction looked like one that could lead in totally new directions, and by the time I finished my walk, I had decided that I would try to solve it.

The disappearance of the Cretaceous forams as recorded in the Scaglia limestones appeared to have taken place suddenly, perhaps even catastrophically. But in the mid-1970s the thought of a catastrophic event in Earth history was disturbing. As a geology student I had learned that catastrophism is unscientific. I had seen how useful the gradualistic view had been to geologists reading the record of Earth history. I had come to honor it as the doctrine of "uniformitarianism" and to avoid any mention of catastrophic events in the Earth's past.

But Nature seemed to be showing us something quite different. That little bed of clay at Gubbio was in conflict with gradualism, the most useful and cherished concept in geology. So let us now see why gradualism had such a tight hold on thinking about Earth history.

CHAPTER 3

Gradualist versus Catastrophist

BIBLICAL CHRONOLOGY
AND CATASTROPHISM

In previous centuries, travelers crossing the Alps on primitive trails faced drowning in wild rivers, freezing in blizzards, or burial by avalanches. As grim obstacles, slashed through by dark canyons and capped by a wilderness of glacial ice, mountains must often have seemed threatening in the past.

When scientists began to turn their attention to what we now call geology, an obvious question was how mountains like the Alps came to be. We now see that the answer to this question depended on how much time was available for their creation. Mountains could form slowly and gradually if there was lots of time in Earth history. However, the early geologists automatically assumed a brief history, because the Bible actually lists the generations of our forefathers back to the creation of the Earth, and the Bible was accepted as an accurate account of history. On this basis, James Ussher, an Anglo-Irish bishop (1581–1656), determined that the Earth had been created in 4004 B.C.

With so little time available for its formation, a mountain range like the Alps could only be seen as the wreckage from a catastrophe, and perhaps this view resonated with the gloom that travelers evidently felt while crossing mountain barriers. As long as the biblical chronology was accepted, people who thought about the history recorded in rocks and landscapes had to conclude that changes in the Earth's past had been very rapid. This viewpoint came to be called catastrophism.

Geology could not become a real science until the stranglehold of Biblical chronology was broken. Geologists have long attributed this breakthrough to two scientific heroes. The first of these was the eighteenth-century Scotsman, James Hutton, who is credited with the discovery that the Earth is enormously ancient. The other was the nineteenth-century Englishman, Charles Lyell, recognized as the father of "uniformitarianism"— the view that all changes in Earth history have been gradual. Although these traditional accounts are now recognized as oversimplified and misleading,[1] they were accepted until recently by most geologists and paleontologists.

TO MAP THE PLANET

Hutton's ancient Earth and Lyell's uniformitarianism gave geologists the tools they needed to approach their central scientific problem—to understand rocks and landscapes. Long-familiar mountains like the Alps and dramatic, newly discovered landscapes like the Grand Canyon no longer required catastrophic explanations. Slow deformation and slow erosion over very long periods of time better explained what geologists saw in the field. John Muir poetically but correctly attributed the vertical walls of Yosemite to the slow grinding of glaciers, rather than to violence and catastrophe: "Nature chose for a tool not the earthquake or lightning to rend and split asunder, not the stormy torrent or eroding rain, but the tender snow-flowers noiselessly falling through unnumbered centuries, the offspring of the sun and sea."[2]

The concept of an ancient Earth made it possible to understand rocks and landscapes correctly, but it raised a new problem. Geologic processes, active through the 4,600 million years of Earth history, have produced an enormously complex and varied array of rocks. Those rocks, constituting the historical record of the Earth, are to be found all around the globe—in farmlands, deserts, mountains, jungles, and under the sea. Describing all those rocks and interpreting all that history would be a formidable task. It would take generations of geologists to complete. And so, beginning in the nineteenth century, geologists settled down to do what was clearly a necessary task—to measure and describe the rocks of the entire surface of the world and to plot their distribution on detailed maps that would be the basis for understanding Earth history.

Constructing an accurate geologic map of an area, showing the locations of all the different kinds of rocks and their geometrical relationships, is a challenging and rewarding task, and geologists became very skilled at mapping. I've made several geologic maps at a variety of scales, and I take pride and pleasure in them.

As the decades went by, systematic geologic mapping paid off in a more and more detailed knowledge of the history of the Earth, region by region. Mapping led to dramatic discoveries, like the recognition by geologists in the Alps that thick sheets of rock have been pushed up and over younger rocks for many kilometers along breaks called thrust faults.[3] There were also enormous economic benefits, for geologic mapping led to the discovery of huge oil reserves and mineral deposits. It is no exaggeration to say that twentieth-century technical and industrial civilization rests on a foundation of natural resources largely found through geologic mapping.

The need was real and the task enormous. But as generation after generation proceeded with the work of mapping, many lost sight of the original goal of fully understanding the planet. Mapping itself became the goal. The desire of many geologists was to find a new, untouched field area where no one had mapped before. As fresh areas became rarer, geologists looked ever more closely at rocks already known.

Geologic mapping was satisfying and useful, but in retrospect most of it seems to me to have been intellectually pretty routine. While early twentieth-century physicists were reading "the thoughts of God," in Einstein's phrase—exploring the majestic curvature of spacetime on the scale of the universe and discovering the weird quantum behavior of the infinitesimally small— geologists labored to reconstruct the paths of ancient rivers and the pattern of lands and seas at various times in the past. Relativity and quantum mechanics were stretching the minds of physicists almost to the breaking point, forcing them into worlds of thought where no one had gone before, and radically changing our entire concept of the universe. Geology simply asked its apprentices to learn the techniques of geologic mapping and to memorize a lot of complicated terminology and then it sent them out to add to the growing knowledge of the rock record of the Earth.

And yet, with hindsight, we can see that the mapping was an investment which is now paying great dividends. Physics could make great discoveries quickly by reducing complicated problems to simpler components because physics investigates the fundamental laws of Nature, which do not change and do not become more complex through time. Geology seeks to understand the Earth, which has evolved over 4,600 million years, accumulating more and more historical complexity in its rock record. That century and a half of mapping the Earth produced the detailed knowledge of the rock record which is now allowing geology to emerge as a mature science, skilled at interpreting historical complexity, and therefore perhaps the discipline best prepared to lead science into the holistic world of the twenty-first century.

THE LURE OF ADVENTURE

When I arrived at Princeton, my fellow graduate students were mapping the rocks all around the Caribbean as part of the research program of Professor Harry Hess, and I was fascinated by their stories of adventures in the field. Eldridge Moores, later to make fundamental discoveries in the plate tectonic revolution and to become the President of the Geological Society of America, told about the harrowing ordeal of trying to drive Professor Hess over the mountainous Southern Peninsula of Haiti. During a midnight downpour, their jeep kept stalling in a river infested with parasites, while voodoo drums were beating ominously around fires in villages on the banks. Jack Lockwood, who was to spend his career helping people avoid the hazards of active volcanoes around the world, was just back from his first field season mapping in the Guajira Peninsula, at the northern tip of South America, and was full of stories about living with the Indians of that remote desert wilderness. That was what I wanted to do!

Professor Hess accepted me to work on his Caribbean Project, and field geology provided all the adventure I had hoped for. After two seasons of mapping in the Guajira Peninsula, a fortunate blind date led to a whirlwind romance with a young graduate student in psychology named Milly, who loved travel and adventure as much as I did. We spent a long honeymoon in the roadless desert of the Guajira, as I constructed a detailed map of that particular part of the world. We slept in the back room of Robertico Barroso's trading post or in hammocks next to a campfire wherever the mapping took us. We spoke Spanish and learned a little Guajiro from our field companion, Lucho Restrepo, played guitar, sang the songs of Colombia and Venezuela, and came to know a way of life that would soon disappear. We learned about the traditional Indian law which ruled the peninsula and about the symbols of the Guajiro clans, and about the ancient feuds and the use of poisoned arrows. We drove our Land Rover across the sands and up the rocky arroyos, filled our barrel with more-or-less drinkable water from windmills, talked and sang all night with Indians and smugglers over bottles of warm beer, and went to town once a month for supplies. So maybe my Ph.D. thesis was not very challenging scientifically, but it filled in another gap in the geological mapping of the world, and it was an unforgettable adventure.

THE PRESENT IS THE KEY TO THE PAST

As students of geology, learning the skills of field mapping, we absorbed the traditional, exclusive focus on slow, gradual processes. We were proud that our discipline had made one fundamental contribution to the edifice of science. That contribution was not the discovery of the antiquity of the Earth, or the theory of evolution, but the principle of uniformitarianism.

One way of expressing uniformitarianism was with the phrase, "The present is the key to the past." That formulation was a bit vague, but it was useful in many situations. It meant that if you wished to understand the sedimentary deposits of an ancient estuary—the layers of mudstone, the rippled sandstones, and the burrows cutting through the beds—you should go and study a modern estuary like San Francisco Bay, wading through the mudflats, measuring ripples in the sands where tidal currents are flowing, and finding out which kinds of clams are digging the burrows.[4] This formulation of uniformitarianism fit the preference for a quiet, gradualistic Earth, because the only processes geologists could directly observe—and live to tell the tale—were calm, quiet ones. Geologists made great progress in interpreting ancient sedimentary deposits of all kinds by following the uniformitarian admonition that the present is the key to the past. It was an excellent research strategy and led to many discoveries.

But uniformitarianism had a second meaning. Accepting that the present is the key to the past, and being aware that no really major catastrophes have happened in recent human history—nothing worse than big earthquakes or volcanic eruptions—geologists believed for more than a century that catastrophes have played no role in Earth history.[5]

NATURA NON FACIT SALTUM

Evolution in the observational sense—that the fossils change as we work our way up through the sedimentary rocks from older to younger—was an inescapable conclusion to paleontologists as far back as William Smith around 1800. Early paleontologists noticed that at certain levels in the stratigraphic record there were sudden, dramatic changes in the kinds of fossils present. They found that these breaks were recognizable over long distances and could be used as boundaries, to divide Earth history into named intervals. The most recent of these great discontinuities in the fossil record was used to define the boundary between the Cretaceous Period and the Tertiary Period.

Lyell, as part of his uniformitarian view of history, believed that the rate of biological change recorded by fossils has always remained the same. The nearly total difference between the fossils at the top of the Cretaceous and those at the base of the Tertiary thus presented a dilemma for Lyell. He was forced to the conclusion that an enormous amount of time had gone unrecorded—that no sediment had been deposited in any place yet discovered during the long interval that would be needed for those great alterations in fossils to unfold at the usual slow rate of change. More time had passed, unrecorded, at the Cretaceous-Tertiary boundary, he argued, than has passed during all the time since then.[6] This remarkable conclusion was a necessary consequence of his belief in uniformity of rate. The Gubbio limestones with the thin bed of clay at the KT boundary were not studied until long after Lyell's death. His view of history would require that the clay bed represent more time than all the hundreds of meters of Tertiary rocks above it. I wonder what Lyell would have thought of that outcrop. . . .

Until the middle of the nineteenth century, no one understood the reason for the evolutionary changes recorded by fossils. The breakthrough came in 1858, when Wallace and Darwin proposed, independently and almost simultaneously, that evolution occurs through the natural selection of favorable traits. This explanation has in large measure endured to the present time, encapsulated in the phrase, "the survival of the fittest." Although Wallace wrote his paper first, Darwin supplied the convincing detailed evidence, bore the brunt of the angry attacks, and usually gets more of the credit for the theory.

There was an intriguing interplay between Darwin and Lyell in the formulation of the theory of evolution. Darwin was strongly influenced by Lyell's *Principles of Geology* on his long voyage around the world on the *H.M.S. Beagle* from 1831 to

1836, while he was making the observations which later led him to the theory.[7] Darwin made Lyellian gradualism a key tenet of the theory of evolution when he came to write his own classic, *The Origin of Species*. The vulnerability of the gradualism which Darwin built into the theory was immediately recognized by T. H. Huxley, the strong supporter who came to be known as "Darwin's bulldog." Huxley wrote to Darwin, "You have loaded yourself with an unnecessary difficulty in adopting *natura non facit saltum* so unreservedly."[8] *Natura non facit saltum*—Nature does not make sudden jumps—nevertheless remained a central tenet of Darwin's theory, and has strongly influenced paleontological thought down to the present time. Nature, it seemed to most Earth scientists, was a calm, well-regulated domain in which catastrophes and irregularities were forbidden.

A CRACK IN THE UNIFORMITARIAN FACADE— THE SPOKANE FLOOD

In one memorable episode, however, the gradualist orthodoxy was threatened by a dissenting opinion. The challenge arose in the 1920s when J Harlen Bretz, of the University of Chicago, described a network of huge, dry channels in the aptly named "scablands" around Spokane, in eastern Washington. The dry channels looked like river valleys with rippled sand bars, but on a gigantic scale. Bretz proposed that they had been scoured out by the waters of an immense, catastrophic flood during glacial times. A catastrophic flood! It must have seemed like the return of biblical catastrophism. The full weight of uniformitarian doctrine came down on Bretz, as geologists who had never even been to Washington contradicted his interpretation, convinced that catastrophism of any kind was unscientific nonsense.[9]

In all fairness, Bretz was unable to say where the sudden flood of water could have come from, so there was indeed reason for doubt. But Joseph Thomas Pardee recognized beach lines, high on the mountainsides above Missoula, Montana, upstream from the scablands, and correctly interpreted them as marking shorelines of a huge lake that had once filled a deep valley temporarily dammed by glacial ice.[10] By the 1940s it was clear from Pardee's studies that the ice dam of Glacial Lake Missoula had broken as the glacier melted back, releasing an enormous flood of water that catastrophically eroded the channels of the scablands.

Nevertheless, uniformitarian dogma blocked acceptance of the Spokane Flood for two more decades until similar scablands were discovered on space-probe images of Mars. Finally, in 1965, a new examination of the evidence in the field by an international group of geologists made it clear that Bretz had been right all along, and at age 83 he received a congratulatory telegram from the people on the trip that said, "We are now all catastrophists."[11]

Despite the occasional crack in the facade, by the mid-twentieth century, most geologists were still uniformitarians, but few geologists had ever read or really understood Lyell. We only learned that he had been the founder of geology and that the established wisdom of our science was that nothing really dramatic—no catastrophes—had ever happened in the planet's past. This was the mind-set of the time, as Milly and I camped in the Guajira Peninsula in the mid-1960s and I prepared the map required of an apprentice geologist.

GEMINI, APOLLO, AND THE SPACE PROBES

I remember several days in August of 1965, during a severe drought in the Guajira—days which were almost surrealistic at the time and seem strangely symbolic in retrospect. It was the time of the yearly fiesta at the trading post of Taparajuín. People came from everywhere and consumed large quantities of roast goat and sheep and the homemade white lightning called *chirinchi*, to the accompaniment of music and dancing. The parish priest arrived from a hundred kilometers away, baptized the new babies, and prayed for rain. The next day it rained a bit, to the delight of everyone—especially the priest. The following evening Milly and I went to a funeral near Taparajuín and saw how the Guajiros mourn, with more roast goat and sheep and more *chirinchi,* weeping around a fresh coffin under the stars. And as these timeless events unfolded, we would occasionally tune in a shortwave radio and listen to the messages from the Gemini V spacecraft, as Gordon Cooper and Charles Conrad passed overhead through the night sky, circling the Earth again and again as NASA prepared to go to the Moon.

It was a jarring contrast, and it seemed symbolic of a major transition in human history. Soon the Guajiro way of life, like so many traditional ways everywhere, would disappear. The drought continued and many of the Indians moved to the city of Maracaibo. Then a huge coal mine was opened and a railroad laid to carry the coal to a harbor built on the coast of the Guajira.

Although it took many years to play out, that jarring contrast was symbolic of a change in geology as well. My science was about to be shaken from bottom to top. The Space Program continued on track, and on Christmas Eve of 1968, Apollo 8 circled the Moon carrying Frank Borman, Jim Lovell, and Bill Anders, the first humans ever to leave the gravitational captivity of the Earth. In a radio broadcast to Earth, they described the moving contrast between the view down to the desolate waste of the Moon and the view back homeward to the blue and green oasis of Earth, and they read the first verses of Genesis—a Christmas present to a country weary with civil unrest and war in Vietnam. "In the beginning God created the heaven and the Earth . . . and God saw that it was good."

Looking down at the Moon, the Apollo 8 astronauts saw uncountable craters. When other astronauts walked on the Moon a few months later, they collected samples of lunar soil, all full of shattered rock fragments, with glass spheres that had been melted by impact, and with tiny craters pitting the spheres. From the study of these samples it became overwhelmingly clear that the craters which cover the Moon are due to the impact of aster-oids and comets—to the kind of catastrophic events forbidden to geological thought by the ruling uniformitarian view. Ralph Baldwin, whose 1949 book, *The Face of the Moon*,[12] was another lonely crack in the uniformitarian facade, had been correct in at-tributing the lunar craters to impact. Before long, unmanned probes to other planets and moons sent back images which made it clear that impact craters are the rule in the solar system, not the exception. In a few years, geology went from a discipline with a single planet to study, to one overwhelmed by data from so many planets and moons that it was hard to remember them all. And most of those bodies were covered with impact craters.

Eugene Shoemaker was the leading scientist in the geologic exploration of the Moon and the unmanned study of most of the solar system. Almost before any other geologist was interested in the Moon, Gene was using his surveyor's telescope to study the Moon at night, in the clear skies of the Colorado Plateau, where he had been sent by the U.S. Geological Survey to do field map-ping. He proved the impact origin of Meteor Crater. He built the Astrogeology Branch of the Geological Survey at Flagstaff. And only a last-minute health problem kept him from being the first geologist to walk on the Moon.

Gene might have expected that uniformitarian gradualism would be swept away in the excitement produced by the geo-logic discovery that rocky planets and moons throughout the solar system are covered by impact craters. Geology should have been transformed from a science that studied a single planet to a science with an abundance of objects to investigate, almost all of them covered with evidence of catastrophic im-pacts. But in a remarkable irony, that transformation did not happen, and instead, uniformitarianism became still stronger. Moons and planets were largely ignored by the vast majority of geologists, because the revolutionary discoveries of the lunar and planetary missions were overwhelmed by a development that was even more exciting, and which was fully uniformi-tarian—by the plate tectonic revolution.[13]

THE UNIFORMITARIAN TRIUMPH
OF PLATE TECTONICS

Up through the 1950s, most geologists believed that the continents had always been just where they are now. *Vertical* uplift and subsidence had always been undeniable because of evidence from ancient sea level, which imprints a reference mark by the deposition of coastal sediments like coral reefs and beach sands. These ancient sea-level markers may now be found high in the mountains or buried deep below the surface, so vertical movements were accepted and were at the heart of the history geologists sought to unravel

But the possibility of major *horizontal* movements was firmly denied, because geologists could rarely find markers to provide evidence for ancient horizontal movements. Challenging the accepted view that continental positions have always been fixed, the German meteorologist Alfred Wegener argued in the 1920s that the matching coasts of South America and Africa do provide a marker for horizontal motions. He argued that these continents must have once been side by side, and that they have subsequently moved apart through continental drift.[14] Wegener reassembled the present continents into a monster continent that he called "Pangaea," meaning "all the lands." In his reconstruction of Pangaea, not only did the shapes of the pieces fit together, but any number of geologic features—ancient glacial deposits, for example—also fit together. It was like matching the picture on a jigsaw puzzle. But most geologists either ignored or ridiculed Wegener and his continental drift. The most valid reason for ignoring him was that Wegener's driving mechanism just had to be wrong—he pictured the drifting continents plowing through the deep solid Earth like ships through the ocean, and geologists rightly refused to believe that this was physically possible.

Most geologists thus believed in fixed continental positions until the late 1960s, when the plate tectonic revolution took the science of geology by storm.[15] New maps of the magnetic reversals imprinted in ocean crust showed conclusively that some oceans are gradually getting wider, and that new ocean crust is generated at mid-ocean ridges where deep mantle rocks rise and cool. Accurate new determinations of earthquake locations showed that other oceans are shrinking as old ocean crust descends back into the deep Earth at places marked by deep trenches near lines of volcanoes. Geophysicists recognized that the Earth's surface is divided up into a number of plates, each of them moving relative to all the others and carrying the continents along with them. It was even possible to calculate past plate motions.[16] Wegener had been right about continental drift, except that continents do not plow through the deep solid Earth; they are carried along as the deep Earth convects—as it slowly turns over like a simmering pot of thick soup with a floating layer on top—to allow its internal heat to escape.

The plate tectonic discoveries electrified the Earth sciences. Suddenly geologists working on one continent had a reason to be interested in other continents, because widely separated continents might formerly have been adjacent. The ocean depths were intensively explored. Continental-margin sediments and the deformed rocks of mountain ranges were reinterpreted in terms of expanding oceans, or of contracting oceans and colliding continents. The fairly routine field of geology was suddenly transformed into perhaps the most dynamic science of the 1970s. The knowledge of the Earth acquired during more than a century of detailed mapping paid off at last, as it guided the formulation of a theory never envisioned while the mapping was being done.

Few geologists who were involved in the plate tectonic revolution will ever forget the exhilaration of all the new discoveries. Almost all geologists were somehow involved, for plate tectonics affects almost every aspect of geology. Plate processes have acted continuously for at least a thousand million years, and their effects have been imprinted in one way or another on almost every rock on Earth. There were research opportunities on all sides, and the new discoveries poured forth for more than a decade, totally absorbing the interest of geologists everywhere.

Many hundreds of geologists went out in the field and looked at rocks with the new vision of plate tectonics, and came back to contribute to the new understanding of the Earth. Meanwhile, in a wonderful irony, the lunar landings—the first geological explorations of another world—were largely ignored. A single geologist, Jack Schmitt, and a few geologically trained astronauts walked on the Moon. Lunar samples and the space-probe images of planets were studied by a small number of scientists, who recognized the overwhelming evidence for comet and asteroid impacts throughout the solar system.

Catastrophic impacts were real and the strict gradualism which denied the reality of impacts was dead, or should have been. But most geologists were not paying attention. We were overwhelmed by the even more exciting discovery of plate tectonics, and the central concept of plate tectonics is gradual change. Oceans so wide that they take hours to cross in jet aircraft have grown by sea-floor spreading over tens of millions of years, at a rate of a few centimeters per year—about the rate at which your fingernails grow. Plate tectonics was the most gradual, uniformitarian theory imaginable. The lunar and planetary evidence for catastrophic impacts was scarcely noticed by a geological community absorbed in the breakthrough of plate tectonics, and all the old uniformitarian prejudices were reinforced by the triumph of the plate tectonic revolution. The death of strict uniformitarianism required by the craters spread all over the solar system would have to wait.

THE UNIFORMITARIAN VIEW
OF DINOSAUR EXTINCTION

Plate tectonics theory guided much of geological research in the mid-1970s. When Bill Lowrie and I went to the Apennines, as recounted in the last chapter, our idea was to collect paleomagnetic data to test a plate-tectonic idea—to test whether the Italian Peninsula had behaved as a rotating microplate. Our plan had not worked out, but instead we found that we could determine the fossil ages of the magnetic reversals which were the key to the timing of sea-floor spreading and plate motion. But when we became familiar with the KT boundary at Gubbio and learned to find it ourselves, I began to think about problems completely unrelated to plate tectonics: Why had there been a mass extinction at the KT boundary? Why had nearly all the species of forams died out? Why had the dinosaurs disappeared?

I had not learned much about the death of the dinosaurs in college or graduate school. Mass extinction was treated as a nonproblem, barely noted in passing. In 1886, Alexander Winchell, Professor of Geology and Paleontology at the University of Michigan, used melodramatic prose to duck the question: "A higher type is now standing at the threshold of being. A knell is sounding the funeral of the reptilian dynasty. The saurian hordes shrink away before the approach of a superior being. After a splendid reign, the dynasty of reptiles crumbles to the ground, and we know it only from the history written in its ruins."[17] Almost 50 years later, the account in a major textbook was only a little less vague: "Whatever the cause, the latest Mesozoic was a time of trial when many of the hosts were 'tried in the balance and found wanting'—wanting in adaptiveness to the new environment."[18]

Accounts had become somewhat more detailed by the mid-1970s, and they reflected the prevailing uniformitarian view of Earth history, attributing the dinosaur extinction to climate changes or to a fall in sea level. Whatever the mechanism might be, the extinction was viewed as gradual, spread over a few million years at least, and therefore not really a very important question. Every species becomes extinct eventually. Extinction is a continual process, paleontologists said, and the dinosaur species all died out, one after another, late in the Cretaceous, and left no descendants. Dinosaurs, in the general opinion, became extinct with a whimper, not with a bang. It did indeed look that way; dinosaur bones are rare, the stratigraphic record is very incomplete, and with few fossils preserved, a sudden extinction appears gradual.[19] *Tyrannosaurus rex*, the most famous of dinosaurs, is known from only a few fossil specimens. Clearly this does not provide enough information to distinguish a sudden extinction from a gradual one.

In 1978, Steve Gartner and John Keany proposed a terrestrial catastrophe to explain the KT foram extinction.[20] They argued that the latest Cretaceous Arctic Ocean had been isolated and diluted by fresh water from rivers. When a passage opened to the rest of the ocean, they suggested, the fresher Arctic water floated out over the ocean surface, quickly poisoning the forams. I always liked this ingenious idea and thought it might well be right, until the evidence for impact began to mount up.

The main problem with the Arctic spillover hypothesis was that its oceanographic mechanism did not offer an explanation for the dinosaur extinction. Of course, in 1978 most paleontologists believed the end of the dinosaurs to have been gradual and unrelated to the abrupt foram extinction.

One dinosaur paleontologist stood alone. Dale Russell, whose Ph.D. was from Berkeley and who worked at the Canadian National Museum of Natural Sciences in Ottawa, had examined the stratigraphic record of the disappearance of dinosaurs in detail and was convinced that it required a sudden extinction. Dale could not imagine a terrestrial event capable of suddenly killing all the dinosaurs, so he suspected an extraterrestrial cause. Starting from a previous suggestion that radiation from a nearby supernova—an exploding star—could kill organisms on Earth,[21] Dale and physicist Wallace Tucker proposed in 1971 that climate changes triggered by a supernova explosion had caused the extinction of the dinosaurs.[22]

It was a catastrophic hypothesis, contradicting all the training and experience of geologists and paleontologists. Dale's colleagues snickered quietly and ignored him. But the time was coming for Dale Russell's view of sudden extinction due to extraterrestrial causes, and soon there would be a flood of research which would sweep away the strict doctrine of uniformitarian geology.

CHAPTER 4

Iridium

GEOLOGY AND PHYSICS

Uniformitarian gradualism provided an excellent framework for answering questions about the Earth. Geologists learned uniformitarianism from their teachers and found that in practice it almost always led to reliable explanations of geologic features. Exceptions like the scablands of eastern Washington, which seemed to require catastrophic causes, were explained away or ignored. Gradualism had become a dogma.

It took me a while to realize that the thin bed of clay at the KT boundary at Gubbio not only raised the question of what had caused the mass extinction, but that it also seemed to contradict the gradualistic mind-set of geologists. The near extinction of forams at Gubbio looked very abrupt. Al Fischer stressed in his talk at Lamont that the dinosaur extinction had happened at the same time. Could *T. rex* have perished in a catastrophic event?

It seemed to me that the clay layer at Gubbio might hold the answer. But from what I could see with a hand lens and a microscope, it looked like a pretty ordinary clay. Most of the beds of Scaglia limestone are separated by thin clay partings, and although the KT layer was a bit thicker, it did not seem particularly unusual. The only reason that the KT clay attracted attention was that the forams were completely different in the beds above and below it.

If there was a clue in the Gubbio KT clay layer, it would probably be something outside the usual experience of geologists. I started talking about the extinction question with my father, Luis W. Alvarez, a physics professor at the University of California at Berkeley. Dad was a master of experimental physics and the leader of a research group at the Lawrence Berkeley

Laboratory that had discovered a whole zoo of subatomic particles—work which had been honored with a Nobel Prize in 1968.
He always had a broad curiosity and the ability to dream up
novel approaches to interesting problems—as in the case where
he and his friend, the Egyptian archaeologist Ahmed Fakhry,
had x-rayed the pyramid of Kephren at Giza, using cosmic-ray
muons.[1] They had hoped to discover unopened rooms full of
treasure, but found instead that the pyramid was solid rock
from bottom to top.

Dad did not originally think that geology was an interesting
science. It was my Mother, Geraldine, who got me interested in
rocks when I was in high school, by taking me and my sister
Jean on train trips through the spectacular scenery of the West,
and by showing me places in the Berkeley Hills to collect minerals. She still likes to remind me that my first rock hammer was
one I borrowed from her—and lost!

I had gone away to college and graduate school, then lived in
Holland, Libya, and Italy. I rarely saw my father, and did not
know much about him as a scientist. But when Milly and I returned from Italy in 1971 and I started working as a researcher
at Lamont, Dad came to visit. He was intrigued by all the geological and geophysical data he saw at Lamont and by the excitement of the plate tectonic revolution that was just cresting,
and we thought it would be interesting to try to combine his
physics and my geology. We talked on the phone many times
and came up with some ingenious techniques for dating rocks.
We thought our ideas were new, but unfortunately other people had previously invented them and found out why they
wouldn't work. Nevertheless, it whetted our appetite for working together.

MEASURING LYELL'S GAP WITH BERYLLIUM

By 1976, I was focusing my attention on the KT
extinction and I began to discuss it with Dad. What specific
question could we ask about the clay bed at Gubbio? What useful measurements could we make?

I pointed out that it would be valuable to know how long it had taken to deposit the clay layer. Very rapid deposition of the clay would suggest a sudden cause for the extinction, but slow deposition would suggest a gradual mechanism. Although we did not know it yet, this was Charles Lyell's old problem in a modern guise. As discussed in chapter 3, Lyell had noted in 1830 that the fossils in the top Cretaceous beds differ more from those of the base Tertiary than the latter differ from living animals. Guided by his gradualistic view, he was forced to conclude that more time lay unrecorded in a stratigraphic gap at the boundary than has passed during the entire Tertiary. By 1976, it was obvious that this was wrong. Radiometric age dates, although not abundant, placed the KT boundary at very roughly 65 million years ago and allowed a gap of no more than a few million years. The enormous faunal change in a short interval of time meant that there had been a mass extinction, and that Lyell's extreme gradual view was wrong.

Paleontologists knew this, but they still had a few million years of leeway and could interpret the KT mass extinction as gradual over that interval. However, Bill Lowrie and I, with Al Fischer's team, had shown that the Scaglia limestone at Gubbio records all the magnetic polarity zones known from the magnetic striping of the ocean basins. The reversed polarity zone that contains the KT boundary seemed to represent about 0.5 million years and was present in the expected thickness; so clearly the KT gap represented less than 0.5 million years, and probably no more than 0.1 million years.

On a *geological* time scale, the mass extinction had been abrupt. But had it been abrupt on a *human* time scale? We needed to measure the sedimentation rate of the KT clay—was it deposited in a year or less, or over millenia? Dad thought of a way to find this out. He suggested that we measure the abundance in the KT clay of the isotope beryllium-10, which has 4 protons and 6 neutrons. Be-10 is radioactive, with a short half-life. It is constantly being created in the atmosphere when high-energy cosmic rays—fast-moving atomic nuclei from far away in the galaxy—slam into oxygen and nitrogen in the air, breaking them up into smaller fragments, including Be-10 nuclei.

These freshly made atoms may be incorporated in sediment like the KT clay at Gubbio. The more time the clay represented, the more Be-10 it would contain. The published half-life of Be-10, 2.5 million years, looked just right—fast enough for atoms pre-dating the KT boundary to have disappeared and not confuse the matter, and slow enough for at least some atoms contemporaneous with the clay to still remain. If we could measure the Be-10 content, assume that the production rate 65 million years ago was the same as today, and correct for radioactive decay since then, it would be possible to calculate how much time the clay represented.

Dad knew who could make the beryllium-10 measurements. He put me in touch with Richard Muller, a young physicist at Berkeley who had done his Ph.D. under Dad's supervision and who had just invented a novel technique for dating rocks and other old materials. Rich had realized that a cyclotron—an atom smasher that accelerates atomic nuclei to very high velocities— could be used as a super-sensitive mass spectrometer to make far better determinations of the abundances of different isotopes of an element than had previously been feasible. Age determinations using an accelerator mass spectrometer have been widely used since then, and Rich's work on this and several other projects was about to be recognized with the MacArthur Prize, the Texas Instruments Prize, and the Waterman Award of the National Science Foundation.

Rich had a trip to New York scheduled, and in December of 1976 he came to visit me at Lamont. He gave a lecture to the Lamont scientists about the potential of accelerator dating. I showed him the warehouse full of ocean-bottom sediment cores from all over the world, and we hiked along the cliffs of the Palisades, overlooking the Hudson River, talking about how to apply physics to geological problems. It was the beginning of an enduring friendship.[2]

Rich returned to Berkeley, and we talked on the phone and exchanged letters full of calculations as we planned the measurement of beryllium-10 in the KT boundary clay. Everything was on track until we learned to our dismay that the published half-life was in error. Be-10 actually decayed much faster than we had thought, with a half-life of only 1.5 millions years. There would be so little left after 65 million years that we had no hope of measuring it. The project was over. Scientific research has many disappointments for every success.

BERKELEY

It had been ten years since I received my Ph.D. Milly and I had lived in South America, Holland, Libya, Italy, and New York. The life of the postdoctoral researcher is exciting but precarious, and we were starting to think about a permanent job. In 1977, a teaching position opened up in the Department of Geology and Geophysics at Berkeley. I applied for it, was interviewed by the Berkeley professors, gave the most important lecture of my life, and was invited to join the faculty. It was a stroke of such good fortune that even now I can scarcely believe it happened.

The near miss with Rich had only increased my determination to solve the mystery of the KT extinction. As soon as I got to Berkeley in the fall of 1977, I started spending time with Dad and Rich, to learn more physics and to get them even more interested in geology. Dad was ready for a new project. His dreams of Pharoah's treasure in the pyramids had faded and the hunt for new subatomic particles had slowed down. Dad was restless, and he was ready to dig into a good mystery.

Shortly after I came to Berkeley, I gave Dad a sample of the KT beds from Gubbio, which I had collected that summer with Terry Engelder, a geologist from Lamont. Dad was hooked. We decided to have another go at the question Rich and I had worked on: How much time was represented by the clay layer that marks the extinction level at Gubbio? How sudden had the extinction been?

To be testable, a scientific question should be clearly formulated, and here were the bits of information on which we could base the formulation: The Scaglia rossa limestone was deposited on the floor of a fairly deep sea and is made of 90–95 percent calcium carbonate. The calcium carbonate comes in part

from the forams, and the rest is from the very much smaller coccoliths, or platelets secreted by floating marine algae, which can
only be seen with a powerful microscope. The other 5–10 percent of the Scaglia is made of fine particles of clay which were
originally delivered to the sea by rivers or winds and then settled to the sea floor, together with the forams and coccoliths.
The one-centimeter bed at the KT boundary was different—it
was made largely of clay. It had no original calcium carbonate,
and thus no forams or coccoliths to give a record of the detailed
history of life during the mass extinction.

In formulating the question of how much time the clay bed
represents, we saw two possibilities. In the first scenario, the
rate of clay deposition would have remained constant and the
limestone deposition would have stopped during the extinction—perhaps because the extinction had left few forams or
algae to produce the calcium carbonate. In this case it would
have taken a few thousand years to deposit the clay bed.

In the second scenario, the deposition of calcium carbonate
would have continued uninterrupted and there would have
been a brief pulse of increased supply of clay to the ocean, perhaps due to more active river erosion or intense storms. In
this case the clay bed might represent only a few years. Which
had been constant—the deposition rate of limestone or that of
clay?

The question was now precisely formulated: Did the clay bed
represent a few years, or a few *thousand* years? It was also formulated in a way that would tell us something interesting about
the extinction event: Were the limestone-producing organisms
out of action for a few thousand years, or had there been a few
years of abnormally rapid clay deposition?

IRIDIUM

How could we answer that question? We needed
something that had been deposited in the Scaglia limestone and
in the clay bed at a constant rate, and then we could calculate

the time represented by the clay bed. The year before, Dad had suggested that we use beryllium-10, formed at a constant rate in the atmosphere. Now he had a new idea of the same general kind—that the deposition rate of meteorite dust would be unchanging. Big meteorites fall very occasionally at random places, but fine meteorite dust from outer space falls constantly, as a very light sprinkling all over the Earth, and if we could measure the amount of meteorite dust in the clay bed and in the normal Scaglia limestone, we would know the time represented by the clay.

But this is rare stuff! You don't realize it when the occasional microscopic grain of meteorite dust settles on your hand or your head, and we knew of no way to extract meteorite dust from ancient sediments and weigh it. But there was a chemical approach. Dad realized that we could analyze the clay for one of the platinum-group elements.[3] These elements are far from abundant in meteorites, but are present in quantities sufficient to measure. The Earth as a whole must have about the same fraction of platinum-group elements as meteorites do, because both came from the swirling cloud of dust and gas which condensed to form the solar system. But the Earth's *crust* and *sediments* have much lower contents of platinum-group elements than meteorites do. This is because these elements are absorbed by iron, and the Earth has an immense iron core where the Earth's allotment of the platinum-group elements must be concentrated. As a result, the sediments at the Earth's surface are strongly depleted in them, to the point where they are barely detectable with the most sensitive techniques.

We reasoned that meteorite dust, slowly accumulating over thousands of years, would be the main source of the platinum-group elements in the Scaglia sediments. If the clay bed had been deposited over a few thousand years, it would have had time to accumulate a detectable amount of platinum-group elements, but if it had been rapidly deposited in a few years it would be essentially free of these elements.

From his knowledge of physics, Dad recognized that neutron activation analysis was the appropriate technique for making the measurements, and when he studied the properties of the six platinum-group elements, it was clear that iridium was the one that would work best. We were in luck, because Frank Asaro also worked at Lawrence Berkeley Laboratory. Frank is a nuclear chemist who developed techniques of neutron activation analysis for studying ancient pottery many years earlier, and we hoped that he could measure iridium. Dad and I went to see Frank.

FRANK ASARO

To understand Frank's analytical work, we need to change our units again, as we did when shifting from years to million-years in chapter 2. When chemists like Frank analyze a rock sample, each element is measured and reported as a fraction of the total sample. For "major elements" the fraction is stated as percent, meaning parts-per-hundred, so a particular rock might contain 5.6 percent iron. In addition to major elements, careful analyses will measure some "trace elements," so rare that they are reported as parts-per-million, or ppm. Really sophisticated analytical techniques like neutron activation analysis can measure trace elements at parts-per-billion levels, or ppb. It takes some mental readjusting to think about trace element concentrations, so here's how I do it: The human population of the Earth is about 5 billion, so every five people make up one ppb—it helps me appreciate what a truly tiny concentration 1 ppb represents.

You might ask why anyone would care about concentrations that low, but geochemists have found that rare trace elements, like rare fingerprints at a crime scene, can reveal the most interesting things about events in the past. We did not know it yet, but this was to be the case with the iridium measurements that Frank could make.

How can scientists like Frank possibly measure ppb concentrations? How does neutron activation analysis work? Suppose that all the 5 billion people in the world were gathered together on some vast plain, and you wanted to find out what fraction had been born on a particular day, during a particular 1-second interval. The answer would be a few parts-per-billion, and it would be extremely difficult to measure, unless. . . . Unless perhaps you could arrange for each of those few people to carry a powerful spotlight, and you could float above the crowd at night in a balloon and count the beams of light.

That's roughly how neutron activation analysis works. The rock sample is put in a nuclear reactor and irradiated with neutrons which are absorbed by atoms in the rock, making some of the atoms unstable so that they decay radioactively. This is the "neutron activation" in neutron activation analysis. When an unstable, activated atom decays, it gives off a gamma ray—a single photon of intense light. The photons coming from each element have a specific, characteristic energy, which is a marker for the presence of that particular element. Frank has a detector that registers a count each time a gamma ray with the energy characteristic of activated iridium, for example, passes through it. This is the "analysis" in neutron activation analysis. Frank analyzes for iridium by counting the beams of light.

The principle is not hard to understand, but doing the analyses is very difficult indeed. There are endless possibilities for making mistakes and getting the wrong answer.[4] There are all sorts of calibrations to be made, instruments can malfunction, and contamination is a major danger at the ppb level.[5] A person has to be almost pathologically careful to do this work well, and Frank is one of the best in the business. I think of him as the intellectual heir of Tycho Brahe, the endlessly painstaking Danish Renaissance nobleman whose precise naked-eye measurements of the positions of planets in the sky allowed Kepler to determine their orbits, which in turn led to Newton's unlocking of the laws of motion and of gravity. Precise in habits and speech, Frank hunts down potential mistakes with the ruthlessness of a counterspy, triple checks everything and then checks it again, and he would be mortified if his numbers ever turned out to be wrong. These characteristics also make him a formidable opponent at cards, and he and his wife Lucille are both Life

Masters in bridge. Perhaps by way of compensation, Frank has one of the messiest desks I've ever seen!

Obviously this kind of finicky work is not for everyone, but the potential payoffs can be enormous. We know today what killed the dinosaurs because of Frank Asaro's ability to make these remarkable measurements.

Frank received us courteously and listened politely to our idea. He immediately told us that he was already in contact with Andrei Sarna-Wojcicki of the U.S. Geological Survey, a geologist who studies ancient volcanic ash layers all over the western United States.[6] Andrei had already recognized the potential of iridium as an indicator of sedimentation rate, and had proposed measuring the iridium in soils, which presumably would come from micrometeorites, as a way of dating the soils. This project was on the back burner, but Frank would not agree to work with us until he had checked with Andrei to be sure the two experiments did not overlap.

Frank did not think he would be able to help us. In 15,000 analyses of pottery shards from archaeological sites, he had rarely detected iridium. Fortunately, Frank found the idea interesting enough that he agreed to analyze a dozen of my samples.[7] I carefully chose the samples, including some from the clay bed, some from the limestone beds immediately above and below, and some from much lower down, for comparison.

THE PARTS-PER-BILLION SURPRISE

I gave the samples to Frank in October of 1977 and for months I heard nothing. Neutron activation analysis is an unavoidably slow technique that involves long obligatory waits, and in addition Frank's equipment had been down, and he had a large backlog of samples to run. Months and months passed, but finally in late June of 1978, I got a call from Dad: Frank had at last completed the analyses and something was seriously wrong. Frank wanted to see us.

Dad and I walked into Frank's lab to see what the problem was, and Frank showed us his results. We had expected about 0.1 ppb of iridium if the clay bed had been deposited slowly, and essentially none if it had been deposited fast. We had never anticipated what Frank actually found—3 ppb of iridium in the portion of the clay bed which did not dissolve in acid. Three ppb was an extremely small amount of iridium to be sure, but it was much more than we could explain by either of our scenarios. Later Frank found that some of the iridium had been carried away during acid treatment of the samples, so the final value was 9 ppb.

Where had all that iridium come from? Possibilities quickly sprang to mind: Could it have come from the supernova that Dale Russell and Wallace Tucker had suggested to explain the dinosaur extinction? Did it come from an impacting asteroid or comet? Or could there be a noncatastrophic explanation? Maybe the iridium was deposited from seawater somehow. Or maybe the Earth had encountered a cloud of interstellar dust and gas. What could possibly explain all that iridium?

DENMARK

Before we put a lot of effort into thinking up hypotheses and testing them, we needed to know whether the iridium anomaly was restricted to the rocks around Gubbio or whether it was a global characteristic of the KT boundary and thus a critical clue to the global mass extinction. We obviously could not immediately test whether the anomaly was truly global, but we thought we should at least analyze for iridium at one or two other KT boundaries far from Italy.

So I went to the library to hunt for another KT site. It is amusing now, when we know of more than a hundred KT sites around the world where the iridium anomaly has been found, to remember how hard it was to locate even one more place to sample. There were almost no reports of continuous stratigraphy across the boundary—which of course reflected Lyell's old idea that the boundary represents a major gap in the record. About the only candidate site was in Denmark, where there was a bed of clay separating chalky limestones of Maastrichtian and Danian age exposed in a cliff called Stevns Klint, south of Copenhagen.

Stevns Klint seemed like our only chance for another KT boundary. One of the blessings of a life in science is the world-wide network of friends and colleagues you can build up. So I called Søren Gregersen, a Danish seismologist I had known at Lamont, and on my way home from fieldwork in Italy, Søren met me at Copenhagen Airport. We drove to Stevns Klint with Danish micropaleontologist Inger Bang, scrambled down the cliff, and came to the clay bed.

It was clear right away that something unpleasant had happened to the Danish sea bottom when that clay was deposited. The rest of the cliff was made of white chalk, a kind of soft lime-stone, which was full of fossils of all kinds, representing a healthy sea floor teeming with life. But the clay bed was black, smelled sulfurous, and had no fossils except for fish bones. During the time interval represented by this "fish clay," the healthy sea bottom had turned into a lifeless, stagnant, oxygen-starved graveyard, where dead fish slowly rotted. Oxygen-starvation deposits like this are not uncommon in the rock record, but usually they represent local conditions. Could the Danish fish clay represent a worldwide disaster in the ocean at the time of the KT extinction? The iridium measurements would tell. Søren and Inger and I collected samples from the clay bed and from the chalk above and below. Frank analyzed them, and in the fish clay he found the anomalous iridium concentration.

We couldn't say, on the basis of two locations in Europe, that the KT iridium anomaly was a worldwide feature, but at least it wasn't a local peculiarity at Gubbio. It was not definitive, but we didn't know of any place else in the world with a complete section across the KT boundary where we could look for iridium. Maybe this was all we would ever find. It was time to think about a global explanation for the anomaly.

The cliff of Stevns Klint, in Denmark near Copenhagen. The sloping lower half of the cliff is the top of the Cretaceous, and the steep upper part is the lowest Tertiary.

OUTSHINING THE GALAXY

There was an explanation ready at hand. Perhaps *T. rex* and all the other victims of the KT extinction had been killed by radiation from a supernova. This had already been discussed in the scientific literature,[8] but as a speculation rather than a conclusion, for no one had ever found any evidence of a KT supernova. Our first idea was that the iridium at Gubbio and Stevns Klint might provide that evidence.

Nothing in our normal lives prepares us for the concept of a supernova—the realization that a star, another sun, can suddenly blow up. Astronomers have never had the geologists' uniformitarian preference for quiet developments, and supernovas are spectacular catastrophes indeed.

Normal stars shine because atomic nuclei of hydrogen at their centers are fusing together to form helium and heavier elements, with the release of prodigious amounts of energy. The energy escapes in the form of photons which ricochet around inside the star, sustaining the pressure that keeps the gravity of the star from shrinking it to a much smaller size. When the photons finally reach the outer surface, they streak away into space as starlight, with some of the light warming planets like Earth.

Stars shine steadily for millions or billions of years, but they die when their fuel is exhausted. As their end approaches, they may gradually dim down and fade out, but some stars die a sudden and cataclysmic death. When the hydrogen fuel is finally exhausted, the sustaining pressure disappears and the star finally collapses—and once the pressure is gone, the collapse can be extremely rapid. All of the suddenly unsupported mass of the star falls toward its center. Some fraction of the star matter bounces back from the pile-up at the center, and the bounce throws this material violently outward into the surrounding space. This stellar explosion is so colossal that the supernova may briefly outshine all the other hundred billion stars in its galaxy![9]

A supernova would be instantly fatal to any life on a planet circling it, but life itself could not exist without supernovas. The primordial universe that emerged from the Big Bang consisted almost entirely of hydrogen and helium. All the other elements have been made by nuclear fusion inside stars and flung into space when those stars exploded. The rocky Earth and our carbon-based bodies are largely composed of the debris of stars that blew up. Supernovas made life possible, but if one occurred in a star in the Sun's neighborhood, it would be a major disaster, with the Earth's surface bathed in dangerous or lethal radiation and the climate seriously disrupted.

Fortunately, a nearby supernova should occur only once in a billion years or so, but unlikely events do happen. A supernova just might explain one truly rare event like the KT extinction. So what Dale Russell and Wallace Tucker suggested in 1971 made lots of sense—a supernova might well have killed the dinosaurs.

It was a suggestion that sounded reasonable to astronomers, who have photographed supernovas, and to physicists, who understand the nuclear processes that make stars explode. The idea was unpalatable to geologists partly because of the uniformitarian tradition, but partly because no geologist had ever seen evidence for an ancient supernova in the rock record. What kind of rock record would a supernova leave?

THE GUBBIO CLAY LAYER
AS A SUPERNOVA RECORDER

There was an iridium anomaly right at the extinction level, in at least two places. Could this be evidence for a supernova? All the elements heavier than helium are made in stars and scattered around by supernova explosions. Iridium was one such element, so maybe our anomalous iridium came from a supernova. How could we test that idea? Dad realized that a nearby supernova would deposit plutonium-244 made by stellar processes, as well as iridium, so we could test the supernova hypothesis by analyzing the KT clay for plutonium-244. This isotope of plutonium is radioactive and decays with a half-life of 83 million years. So many half-lives have passed since the formation of the Earth that any primordial plutonium-244 would have decayed away.[10] But only a little less than one half-life has passed since the KT boundary 65 million years ago, so plutonium-244 could be detected with neutron activation analysis if it had been deposited in the KT boundary by a supernova close to the Earth.

By this time, Frank's colleague Helen V. Michel[11] had joined our little group to help Frank with the analytical work. Helen is a skilled plutonium chemist and she was the leader in making our new measurements on the KT clay. Analyzing for plutonium-244 by neutron activation is much more stressful than doing iridium, because you have to work nonstop, making chemical separations and counting the gamma rays before they are all gone. So Helen and Frank worked all day and then all night, with Dad and Milly and me bringing them coffee and donuts. As the light of dawn was gathering outside, the results were finally ready, and. . . .

And there was plutonium-244 in the KT boundary clay! Dad and I were nearly jumping up and down with excitement—a nearby supernova had killed the dinosaurs! It was a really major discovery. Helen and Frank were too tired to do more than nod with satisfaction.

How do you handle a bombshell discovery like this? Dad was ready to announce it right away. I was nervous about jumping the gun. Frank and I went to see Earl Hyde, a nuclear chemist like Frank, who was Deputy Director of the Lawrence Berkeley Laboratory, and we asked him for advice. Earl listened to Frank's detailed account of the entire experimental procedure and then gave us the best advice we ever got: "Do it all over again," he said. "Repeat every single step from the very beginning, on a fresh sample, to be absolutely sure there really is plutonium-244 in that clay."

Another painstakingly careful chemical preparation of rock samples. Another irradiation. Another all-day, all-night session in the lab. More coffee and donuts. Another dawn, and finally the new results. There was shock and bitter disappointment as we stared at the numbers in disbelief. There was absolutely no trace of plutonium-244 in the clay this time. None whatsoever. A careful analysis of the experiment made it clear that the supernova hypothesis was dead. In neutron activation analysis, if one measurement on a sample detects an element but another does not, the latter is correct and the first run must have contained an impurity, because there is no way to miss those beams of gamma rays if the element is really present.[12] We went home in the early morning dejected, but Earl Hyde had saved us from the humiliation of having to retract a spectacular mistake.

IMPACT?

A supernova was out. What else might have caused the KT extinction? We had to keep reminding ourselves that the iridium anomaly coincided with evidence for the near extinction of planktic forams, nothing more. But it was hard to avoid thinking of the event as the *dinosaur* extinction as well, for the sudden nature of the foram event inclined us toward Dale Russell's heretical view that the dinosaurs also had died out suddenly. From a review of the paleontological literature, Dale estimated that almost half the genera of animals, plants, and single-celled organisms had died out at the end of the Cretaceous.[13] Soon Dale's estimate would be superseded by a detailed database of fossil ranges for marine invertebrates from the literature compiled by Jack Sepkoski of the University of Chicago and analyzed by Sepkoski and his colleague David Raup.[14] The Raup-Sepkoski work showed very clearly that the KT event had affected a wide range of organisms and that the extinctions were synchronous or nearly so.

So why be timid? We began to formulate the question in this way: "What extraterrestrial event could have caused the sudden extinction of half the genera on Earth, while depositing the telltale iridium anomaly?" This was to get us in trouble with many paleontologists, who did not think a geologist, a physicist, and two nuclear chemists should be trespassing in someone else's area of science. Other paleontologists, like Steve Gould, Jack Sepkoski, and Dave Raup, welcomed this input and began to explore the possible implications for paleontology.

We had focused first on the supernova hypothesis because it had been discussed already by Russell and Tucker, and by physicist Malvin A. Ruderman, who was an old friend of Dad's, but there was another possibility. Maybe the extraterrestrial event responsible for the iridium and the extinction had been a giant impact.

Looking back on it, I can no longer remember when the idea of a KT impact first came up. Even as a geologist trained in uniformitarianism and working on uniformitarian plate tectonics, I was aware of the tiny community of lunar and planetary geologists who were interested in craters. I had once been invited to a conference on planetary geology to talk about Italian volcanoes as an analogue to Martian volcanoes, and at the conference there were many talks about impact craters on the Moon and the planets. At another meeting I had been intrigued by a presentation by Robert Dietz,[15] showing a map of craters on the Earth which Dietz and Gene Shoemaker and a few others attributed to impacts.

Pioneering impact geologists like Shoemaker and Dietz were largely ignored. After the lunar landings there was no longer much objection to attributing the craters on the Moon to impact, but most of those craters were obviously very old. Few geologists accepted impact on the Earth as a significant process at any time in the last 500 million years—the time of abundant fossils which allow a detailed understanding of Earth history. There are of course many craters on the Earth's surface, but almost all of them are products of volcanic eruptions. Dietz and Shoemaker were talking about the occasional crater that is not associated with any volcanic rocks. The finest example is Meteor Crater in Arizona, where Gene had found really convincing evidence for an impact origin.[16] Conventional geologic opinion attributed these craters to mysterious explosions that occurred at random times and places for no evident reason.[17] In retrospect this causeless mechanism for making craters is indistinguishable from magic, but at the time many geologists considered it preferable to catastrophic impacts.

I believe that as Dad, Frank, Helen, and I tried to make sense of the iridium anomaly, we sometimes talked about a giant impact, but could not understand why an impact would cause worldwide extinction. Of course the blast would wipe out the nearby fauna, but farther away the animals would survive and would quickly repopulate the devastated area. Impact in the ocean would cause a giant tsunami, but such a tsunami would be confined to a single ocean, and the effects would not be worldwide. A supernova had seemed more reasonable because it would have bathed the entire Earth in lethal radiation, thus explaining the global character of the extinction. But a supernova was out, and impact seemed to provide no global killing mechanism. For over a year we had searching discussions that always ended in frustration, and I would lie awake at night thinking, "There just *has* to be a connection between the extinction and the iridium. What can it possibly be?"

During the summer of 1979, while I was doing paleomagnetic studies in the Apennines, Dad worked hard at finding a global killing mechanism. Day after day he would come up with scenarios and would try them out on Frank Asaro, Rich Muller, and another of his young colleagues, Andy Buffington. Every scenario had some flaw and had to be rejected.

Dad spent a lot of time talking with Berkeley astronomy professor Chris McKee that summer, and it was Chris who got Dad to take the impact idea seriously. Finally Dad started thinking about the dust that would be thrown into the air by an impact. He remembered reading that the 1883 explosion of the Indonesian volcano, Krakatoa, had blown so much dust and ash into the atmosphere that brightly colored sunsets were seen for months in London, on the other side of the world, and he tracked down the book he remembered.[18] Scale the Krakatoa event up to the size of a giant impact, thought Dad, and there would be so much dust in the air that it would get dark all around the world. With no sunlight, plants would stop growing, the whole food chain would collapse, and the result would be a mass extinction. It was the first good hypothesis for why a large impact would cause a global mass extinction. Dad tried as hard as he could to find something wrong with his dust-and-darkness scenario. He calculated as well as he could how much dust there would be, and how dark it would get. Frank, Rich, and Andy could find nothing wrong with the calculations, and neither could Dad. With mounting excitement he got on the phone and called me in Italy.

BACK TO DENMARK

"We've got the answer!" Dad told me, "You have to present it in Denmark." There was to be a big meeting on the KT boundary extinction in Copenhagen in September—an unusual sign of interest in a problem few people cared about. I was going there at the end of my field season, to present the iridium data and the negative results of our test of the supernova hypothesis. Dad now strongly favored the view that darkness due to the dust from a big impact had caused the extinction, and he was sure that everyone at the Copenhagen meeting would be delighted to know why the dinosaurs had disappeared.

I knew geologists and paleontologists better than Dad did, and I was pretty sure there would be strong resistance, and even hostility, to a nonuniformitarian explanation. What's more, I had not been in on developing and critiquing the impact-dust scenario, and was not immediately convinced myself. I remembered well how we had nearly made a bad mistake with the supernova hypothesis—one close call was enough! I told Dad I would go ahead with our original plan to present the iridium anomaly and show that it was not due to a supernova, and that we should carefully evaluate the evidence for impact and dust after I got back to Berkeley.

The Copenhagen meeting would be a major test. We had given short talks about the iridium before and these had been reported in the press, but we had not yet presented the results in front of a knowledgeable audience. Copenhagen would be full of people who really knew the KT boundary. How would they react to our iridium anomaly? As I got off the airplane in Copenhagen, I sensed that a debate was about to begin. I could not have guessed how big a debate it would become.

JAN SMIT AT COPENHAGEN

Standing in line for lunch on the first day of the meeting, I found myself next to a tall, blond young man who introduced himself, in a pleasant Dutch accent, as Jan Smit, from Amsterdam. Jan said to me, "I read a story about your iridium anomaly in the *New Scientist,* and I want to tell you that I've confirmed your discovery. I have a really complete KT boundary section at Caravaca, in Spain, and it has anomalous iridium, too!" It was further evidence of the global nature of the iridium anomaly, and it was the beginning of a deep friendship which would carry us together through 15 years of intense intellectual controversy.

It would be some years before I fully understood the degree of personal integrity that lay behind Jan's opening remark. Studying the rock record of southern Spain for his Ph.D. thesis, Jan had been intrigued by the abrupt KT extinction of forams at Caravaca, just as I had at Gubbio. Looking for a chemical clue to the KT event, he had contacted Belgian neutron activation analyst Jan Hertogen, just as we had contacted Frank Asaro at Berkeley. Hertogen had found high iridium values, but at the time Jan was sick with mononucleosis and not up to looking at the chemical data. As he was recovering he came across the article about our work, looked for iridium in the data printouts, and there was the immediate confirmation.

Some scientists might have been tempted to claim an independent discovery or quickly rush out a paper to establish priority of publication. But from the moment we met, Jan treated his analyses as a confirmation of our discovery. This is the high standard of ethical behavior that scientists aspire to, and which makes the collaborative scientific endeavor possible, but which is not always met because scientists are very human. I hope I would have had the character to do as Jan did, if the roles had been reversed. Now that I know the whole story, I have come to consider Jan Smit the codiscoverer of the evidence for impact.

So Jan and I were convinced that our iridium was evidence for a major extraterrestrial catastrophe of some sort, but as we listened to the uniformitarian views at Copenhagen we began to realize that persuading other geologists and paleontologists would require detailed evidence and extensive debate.

PUBLICATION

Back in Berkeley, our little group dug into the task of developing and testing the impact idea and writing it up for publication. We were under increasing pressure, because the

The Berkeley group in the neutron activation laboratory about 1980.
Left to right: Luis Alvarez, Walter Alvarez, Frank Asaro, Helen Michel.

iridium anomaly was being talked about and other labs were
starting to analyze KT boundary sediments for iridium. We
could find no serious contradictions in the impact hypothesis,
and at the last minute Dale Russell sent us KT samples he had
collected in New Zealand, and they showed the iridium anomaly as well. Finally, in June of 1980, our paper came out in the
journal *Science* and the iridium anomaly was formally established in the scientific literature.[19] Almost immediately there
were three other documentations of KT boundary iridium
anomalies. Smit and Hertogen reported their iridium anomaly
from Caravaca.[20] Frank Kyte, Zhiming Zhou, and John Wasson
at UCLA confirmed the Stevns Klint anomaly and found a new
one in a deep Pacific core,[21] and R. Ganapathy at the Baker
Chemical Company confirmed the Stevns Klint anomaly as
well.[22]

All of these sites were in marine sedimentary rocks, and some
skeptics suggested that the iridium had come out of the seawater. But by the next year, Carl Orth of Los Alamos found irid-

Chuck Pillmore pointing to the thin white band of KT crater debris deposited above sea level that he discovered at Clear Creek, in the Raton Basin of Colorado and New Mexico.

ium in a clay layer in nonmarine coal-swamp rocks in New Mexico, Bob Tschudy showed from pollen studies that it was really of KT age, and Chuck Pillmore of the U.S. Geological Survey found several other KT outcrops nearby.[23] The coal-swamp anomaly showed that the iridium did not come from the ocean, and the case for impact was strengthened.

The matter was by no means settled, however. Scientific hypotheses are tested in the crucible of intensely skeptical criticism. The crucible was warming up, and the impact hypothesis would be tested severely indeed.

CHAPTER 5

The Search for the Impact Site

The Copenhagen meeting in September of 1979 and the iridium papers of 1980 triggered a storm over the Cretaceous-Tertiary mass extinction that raged through the entire decade of the 1980s. Those of us who were involved felt like we were detectives trying to solve a difficult mystery. But the crime had happened so long ago that the trail of evidence had grown very cold. As we struggled to understand what had happened, it almost seemed as if Nature had cleverly constructed a maze of alibis, misleading clues, and false trails.

THE DETECTIVES GATHER

Scientists cannot resist a good mystery. Now that the iridium anomaly was clearly real and probably global, the impact hypothesis attracted hundreds of scientists, who dropped whatever they were doing and started to look for new evidence bearing on the extinction event. In the decade of the 1980s, over 2,000 scientific papers were published on this topic,[1] and it got to the point where there were surprise discoveries almost every month.

Rarely has a scientific question drawn in people from so many completely different disciplines. Geologists and paleontologists were central from the beginning, because it was a problem in reading Earth history and a challenge to the doctrine of uniformitarianism. Analytical chemists, mineralogists, and geochemists joined in to analyze the boundary layer and interpret the subtle chemical evidence. Astronomers found that their understanding of comets, asteroids, and orbital dynamics was of critical importance.

Physicists were drawn in because the instantaneous release of energy equivalent to 10,000 times the world's nuclear arsenal would produce conditions that could never be reproduced in the laboratory and for which there were no adequate computational methods. Atmospheric scientists calculated the physical and chemical effects of a large impact on the chemistry and circulation of the air. Paleoecologists looked for patterns in victims and survivors which might clarify the killing mechanisms. Statisticians probed the question of what inferences could reliably be drawn from very incomplete paleontological data.

Each of these disciplines has its own traditions, its own body of knowledge, and its own specialized language, and these differences raise barriers that normally prevent specialists from working together across discipline boundaries. Had we let these barriers prevail, little progress would have been made in understanding the KT extinction.

Two men quickly recognized the interdisciplinary character of the work which was to come and the need to bridge the inevitable communications gap. Lee Hunt and Lee Silver gathered together a small group of colleagues from various disciplines,[2] and they organized a meeting at Snowbird, Utah, in 1981, in the fall when there were no snowfields and no skiers. The group specifically designed their meeting to teach us all to communicate with each other, and they set up tutorials in which paleontologists learned about the physics of impacts, and astronomers learned about reading the rock record.[3]

That first Snowbird meeting gave birth to a unique scientific culture, in which a specialist in one field is not afraid to ask the most basic questions about a remote discipline, and no one hides behind the obscure jargon that specialists so often use to exclude outsiders. Interdisciplinary conversations became a particular pleasure of this research, and we came to understand and enjoy the very different folkways and languages of each scientific tradition.[4] Physicist Rich Muller, nuclear chemist Frank Asaro, and astronomer Dave Cudaback all came to my Italian headquarters in the town of Piobbico to work in the field with me and learn geology. Dad also visited Piobbico and Gubbio,

Rich Muller and Walter Alvarez at the KT boundary outcrop at Gubbio. The hammer rests on the top bed of the Cretaceous, and the overlying boundary clay is in shadow because many geologists have dug back to collect samples. The darker beds in the upper left are the first deposits of the Tertiary. The limestone beds here have been tilted to the left by the deformation that produced the Apennine Mountains.

with my stepmother, Jan, brother Don, and sister Helen, to see in the field the rocks which had held his interest for so long in the lab. In return, I spent time with them to learn astronomy, chemistry, and physics. Interdisciplinary research groups sprang up all over the world. What might otherwise have been considered scientific trespassing became the expected thing to do. I don't know any other field of research in which so many disciplines work so well together.

It would be misleading to represent the Cretaceous-Tertiary debate as always well mannered and friendly. The ingrained uniformitarian foundation of geology and paleontology was under assault. Strongly held opinions were being challenged on all sides and new information was forcing most of us to revise our understanding and our published views again and again. The very different traditions and methods of different sciences were forced to coexist, and sometimes people on all sides made remarks they later regretted. The effect of an ill-chosen comment was amplified by the fact that the public was interested and the press was closely following the developments.[5] Journalists thrive on hostile confrontations, whereas scientists benefit from intense but mutually respectful debate. We did not all handle the provocations perfectly, and in a few cases serious offense was taken, but I think the field as a whole did reasonably well in maintaining a civilized level of discourse.

POSTMORTEM ON AN INVESTIGATION

The story of research on the Cretaceous-Tertiary extinction through the 1980s is complicated, because so many people played a part and so many scientific disciplines and kinds of evidence were involved. Anyone preparing to recount the events has to choose a way of organizing the material and deciding what to include and what to exclude.[6] The story has been told several times,[7] and it has usually been presented as a conflict between those convinced by the evidence for impact and those arguing the case for volcanism as the cause of the extinction. I prefer to tell it in a different way. I want to focus on the search for the crater which must have been excavated if the impact hypothesis was right, and to consider why finding that crater was so difficult.

As scientists, we are engaged in a conversation with Nature. We ask questions—like "Where is the crater?"—by making observations or performing experiments. And Nature answers, with the results of the observation or the experiment. It seems a straightforward thing to do, but in practice it is very difficult. A young scientist, just starting out, cannot imagine how hard it is to understand the real meaning of Nature's answers, or how many ways there are to make mistakes and get fooled.

Why was it so hard to find the crater? Looking back on the research of the 1980s, it almost seems as if Nature was a skilled mystery writer, setting up a series of clues to be as misleading as possible. Now that the crater has been found, it is too late to tell the story as a suspenseful murder mystery, so let me tell it instead as a kind of postmortem on the solving of a mystery. I will try to tease apart the main threads of investigation, in roughly chronological order, noting how frequently we drew the wrong inference or went off in the wrong direction. It is a salutary lesson in humility!

INITIAL DOUBTS—WAS THERE REALLY A SUDDEN EXTINCTION TO EXPLAIN?

From the very beginning of the search for the crater, there were scientific detectives who argued that we were all on the wrong track—that there had been no crime at all! In their view, the dinosaurs had died out gradually, through natural causes as it were, and even if there had been an impact, it had nothing to do with the dinosaur extinction.

This view was widely held by those who knew the most about the fossil record—by paleontologists, particularly ones who specialized in dinosaurs and mammals. Prominent among these skeptics was, and still is, my Berkeley colleague Bill Clemens.[8] Bill and his students had been working for years in eastern Montana, which is the best place in the world—perhaps the only place—where a stratigraphic record of the very end of the dinosaur era is preserved. This record is not easy to interpret. The last dinosaurs were living on a floodplain crossed by rivers that cut meandering channels that silted up as the river bends migrated across the plain. The channels make it tricky to work out the sequence of events, and the problem in reading the record is compounded by the scarcity of dinosaur fossils. Large living animals are rarer than small ones, and the same is true of fossil remains. One can pick up lots of bone fragments while walking across Bill's research area, but these fragments could easily have been moved by the ancient rivers and do not necessarily indicate that dinosaurs lived at the time the enclosing sediment was deposited. Only articulated skeletons, with the bones still lying together in their original position, provide reliable evidence for the stratigraphic range of dinosaurs, and these are frustratingly rare.

Bill knew roughly where the dinosaur extinction level was. He carefully collected samples and gave them to Frank and Helen, who found the iridium anomaly in a bed from a little nameless butte which came to be called Iridium Hill. The highest dinosaur bone was about 3 meters lower down in the stratigraphy. For Bill, this indicated that the dinosaurs were all dead before the impact that deposited the iridium.[9]

A big debate ensued, dealing with preservation of bone material, and with full skeletons versus bone scrap. You could never realistically hope to find the remains of a dinosaur that was alive at the time of impact, and in fact you should *expect* a substantial gap between the highest fossil and the extinction level. Phil Signor and Jere Lipps at the University of California at Davis made a detailed analysis showing that an abrupt extinction will look gradual if only a few fossils are preserved, and this became known as the Signor-Lipps effect.[10]

One good set of connected bones *above* the KT boundary would have offered strong evidence that the impact did not kill the dinosaurs, but none have been found. On the other hand, the Signor-Lipps effect predicted that as more and more fossils were discovered, the apparent gap between the highest one and the extinction level would shrink, and indeed the original four-meter gap has been reduced to less than one meter. For me that reduction supports the view that the extinction was abrupt and coincided with the impact, but the remaining gap still makes Bill doubt that an impact killed the dinosaurs.

Meanwhile in New Mexico, where for some reason no dinosaur bones at all are preserved, Chuck Pillmore has found a dinosaur footprint less than a meter below the KT boundary iridium level, and no footprints above the boundary. This is just what the impact theory for the extinction would predict.

Another footprint that Chuck found below the boundary was identified by track expert Martin Lockley as the first known footprint of *T. rex*. Because it is hard to be absolutely sure which animal made a particular footprint, track fossils are given slightly different genus names, and completely different species names, from the animal presumed to have made them. Lockley named this footprint after its discoverer, so if you were eaten by one, it was *Tyrannosaurus rex*, but if you were stepped on by one, it was *Tyrannosauripus pillmorei*. What an honor for Chuck!

Smaller fossils are more abundant, which makes it possible to pinpoint their extinction level in the fossil record more closely than is the case for dinosaurs. The best known marine invertebrates that died out at the KT boundary are the ammonites—extinct relatives of the present-day chambered nautilus. Peter Ward at the University of Washington is the leading expert on these fossils,[11] and at first it appeared to him that they had died out before the KT boundary. But after exhaustive collecting in spectacular coastal outcrops in northern Spain, Peter was able to fill in the gap with ammonites right up to the boundary, and he now attributes their extinction to the KT impact. However, another important group of Cretaceous invertebrates that Peter studies, the inoceramids, *do* seem to have become extinct well before the boundary. The history of evolution is complex—neither entirely gradual nor completely catastrophic—and there is much still to do in fully understanding that history.

Even smaller and more abundant are fossils of the single-celled foraminifera in marine strata, and the pollen found in land sediments. There are so many of these tiny fossils that statistical problems like the Signor-Lipps effect do not arise. In the Gubbio and Caravaca limestone there are thousands of forams in every piece of rock you pick up, and the near extinction of these microfossils falls within millimeters of the iridium that resulted from the impact. In the land sediments of New Mexico, abundant pollen reveals a sudden extinction of some plants exactly at the iridium level and a sudden profusion of ferns immediately above it, showing that these disaster-resistant plants flourished in the devastated landscape after the impact.[12]

It seems to me, and to many paleontologists and stratigraphers who are experts in these matters, that the fossil record is most reasonably interpreted as showing that there was indeed an abrupt mass extinction precisely at the end of the Cretaceous.[13] Nevertheless, other knowledgeable paleontologists who are actively studying the fossil record of the KT event continue to argue strongly that the extinctions were gradual.[14] Maybe this disagreement is not surprising. It is hard to read the details of the fossil record, and the details will be critical for a full understanding of the extinction event.

Despite the initial paleontological doubts, the KT iridium anomaly convinced some of us, from 1980 on, that it was worth searching for a giant impact crater of that age. No such crater had been found. Where could it be?

RICHARD GRIEVE'S LIST OF CRATERS

Fewer than a hundred impact craters had been identified by 1980. Soviet and Canadian geologists were the most successful in the search for craters, because their countries contain large tracts exposing very old rocks with a higher probability of having been hit by impacting objects over long stretches of time. Canadian geologist Richard Greive had compiled a list of authenticated impact craters,[15] and many of us studied the list with great care, looking for one that might be the KT impact site. Dating ancient impact craters is difficult, and the ages of many craters on the list were very uncertain. For the most part, Grieve's craters were too small—a few tens of kilometers at most, whereas we estimated that the KT crater should be 150–200 km in diameter. Only three craters in the list approached that size, and those at least were clearly the wrong age. It seemed unlikely that the KT crater had already been found. Why are there so few impact craters on the Earth?

The Moon is peppered with impact craters, but Earth is not. The craters on the Moon are mostly very ancient, dating from the early history of the solar system, when the growing planets and moons were undergoing intense bombardment. The Moon is so small that it quickly lost its internal heat, its water, and its air. It has been inert and inactive for so long that it still preserves its original, heavily scarred surface as a kind of museum of early solar-system history.

By contrast, the much larger Earth is still hot inside, and its interior continually rolls over in the slow convection that drives plate tectonics. In addition, Earth has ice and water and atmosphere which move around and interact continuously, eroding the bedrock in some areas and covering it with sediments elsewhere. As a result, none of the original, heavily impacted surface has been preserved on the Earth. The few impact craters on Earth date from more recent times, after the solar system debris had been mostly swept up and impacts had become small and infrequent.

Where could the KT impact site be? A crater that big would have shattered the impact site for 30 or 40 km down into the crust and the underlying mantle, and it would not be possible to erode away the evidence that far down. It seemed unlikely that such a big impact site was exposed to view and had remained undetected, so there were three possibilities—(a) the crater was covered over by younger sediments or by the ice on Greenland or Antarctica, (b) it was submerged in the ocean, or (c) it had been destroyed by plate-tectonic subduction of oceanic crust.

THE SCENE OF THE CRIME—CONTINENT OR OCEAN?

The obvious first question was this: Did the impact take place on a continent or in the ocean? To a geologist, continent vs. ocean basin is the fundamental dichotomy in the Earth's crust, and the two are not different simply because one is above sea level and the other below. Continental and oceanic crust are different in their chemical composition and in the minerals of which their rocks are composed.[16]

Another key difference is that continental crust is permanent, although continental fragments may split apart and recombine through continental drift. But oceans are ephemeral, with new oceanic crust forming from the mantle between continents that are separating, and old oceanic crust eventually sinking back into the mantle. As a result, there is no preserved oceanic crust older than about 180 million years. Continent and ocean are so different that large impacts on the one or the other should produce very different debris and very different consequences.

Geologists can learn a great deal from the chemical compositions of rocks. In the previous chapter we saw how Frank Asaro's analyses of the trace element iridium at the parts-per-billion level provided the first evidence for the KT impact. Let us now see how the major elements, which occur at the percent level, provided evidence for the location of the KT crater, although we misinterpreted that evidence for years.

The minerals that are important in the Earth's crust, both continental and oceanic, are made of large, negatively charged oxygen atoms held together by a variety of mostly smaller, positively charged atoms, of which silicon is the most important. We use the term "silicates" for minerals based on silicon and oxygen. The simplest of the silicates is quartz, which has two oxygens for every silicon, so that its chemical formula is SiO_2.[17]

Continental-crust rocks are dominated by quartz and by two other silicate minerals called feldspars, which contain aluminum (Al), sodium (Na), and potassium (K). The minerals of oceanic crust are silicates in which calcium (Ca) and magnesium (Mg) are important. Prominent in the oceanic crust are olivine, pyroxene, and calcium-rich feldspar. The oceanic minerals are denser than the continental ones, which explains the difference in height between ocean basin and continent. Here is a very simplified summary:

Continental-crust minerals and their chemical composition:

Quartz	SiO_2
Potassium feldspar	$KAlSi_3O_8$
Sodium feldspar	$NaAlSi_3O_8$

Oceanic-crust minerals and their chemical composition:

Olivine	Mg_2SiO_4
Pyroxene	$CaMgSi_2O_6$
Calcium feldspar	$CaAl_2Si_2O_8$

From this summary it is clear that K and Na characterize continental crust, whereas Ca and Mg are markers for oceanic crust.

EVIDENCE FOR AN OCEANIC IMPACT

Since the rocks of continents and oceans are different in their chemistry, it might be possible to place the KT impact site on one or the other, if someone could find some of the actual debris from the target rock—from the bedrock in the place where the impactor hit. Jan Smit was the first to discover target-rock debris. Studying his Spanish KT boundary samples from Caravaca, Jan noticed some sand-grain-sized, rounded white objects, of peculiar composition, which he called spherules. The spherules held a clue to the location of the crater—continent or ocean—but it was such a subtle clue that no one fully understood the spherules for many years, and even now they still hold some mysteries.

Using the geologist's method for studying rocks and minerals, Jan cut the spherules in half, glued them to a piece of glass, and ground them so thin that they became transparent. Studying these thin sections with a microscope, he saw that the internal crystal structure was feathery—a very strange shape for a mineral grain. When he analyzed them chemically with the electron microprobe, he found that the feathery crystals were made of the mineral sanidine, a kind of potassium feldspar, and a very strange mineral to find in a sedimentary rock.[18]

Jan had gone to UCLA to work with Frank Kyte and John Wasson, and he joined a team led by Don DePaolo to study the isotope geochemistry of the boundary layer. Since World War II, the study of isotopes[19] has yielded a cornucopia of information about all aspects of the Earth, and DePaolo was one of the brightest young figures in this field of science. He has since come to Berkeley and built a major isotope laboratory.

The DePaolo group showed how you could separate Jan Smit's Spanish KT boundary clay into components with four different origins: (1) impactor, (2) target rock, (3) local Spanish sediment, and (4) later replacements. In a beautiful analysis of the isotopes of strontium and neodymium, they showed that the target rock component in the clay layer was completely different from continental crust, but matched oceanic crust very well. It seemed clear from their study that the impact had been in the ocean.[20]

They were also able to show, using oxygen-isotope ratios, that the sanidine in Jan's spherules was not original—it came neither from the impactor nor from the target, but instead was a replacement mineral that had grown later. The original minerals of the spherules had been something different. The identity of the original minerals would be recognized by a new contributor to KT research—Alessandro Montanari, known to all as Sandro.

I met Sandro in the summer of 1978, completely by accident, high in the Apennine Mountains near Gubbio, while he was working on his bachelor's degree at Urbino. We talked about Apennine geology as we ate lunch overlooking a rocky gorge, had dinner in a little village, and started playing music together. Sandro applied to Berkeley and was accepted as a graduate student.

Sandro and I tracked down and sampled many KT boundaries in Italy, and Sandro found that they contained spherules much like Jan's. When he studied the spherules under the microscope, he found more of the unusual crystal textures—in some places like the branches on a snowflake, and elsewhere with starbursts of radiating fibers. My colleague on the Berkeley faculty, Dick Hay, recognized that these were the textures of olivine, pyroxene, and calcium-rich feldspar crystallized from molten rock at an unusual, intermediate rate of cooling—neither the slow cooling that produces full crystals nor the quick cooling that produces glass. Dick showed us articles which described the snowflake and radiating textures in impact spherules brought back from the Moon. Our team, led by Sandro, thus inferred that the original minerals in the KT spherules had been olivine, pyroxene, and calcium-rich feldspar.[21] This was supported a few years later when Jan found one KT site, a drill hole in the sediments on the floor of the Pacific Ocean, where the feathery crystals in the spherules were not altered, and indeed they were made of pyroxene.

Olivine, pyroxene, and calcium-rich feldspar! These are the characteristic minerals of basalt—the main rock of the ocean crust. Everything fit together, with Sandro's group confirming the conclusions of Don DePaolo's group. We could only conclude that the spherules were the result of impact on oceanic crust.

But we had all been fooled! We had drawn the obvious conclusion from the chemical, mineralogical, and isotopic evidence, and as a result, a great deal of effort would go into looking for the KT crater in the oceans, when the crater was really on a continent.

How did Nature fool us? Only years later, after the Yucatán crater was finally found, did we come to understand how we had been misled. The Yucatán Peninsula has continental crust at depth, but it is overlain by a thick layer of sedimentary rock that was deposited on top of the slowly subsiding continental crust. The main minerals in the sedimentary rocks were calcite, dolomite, and anhydrite—minerals based on carbon (C) and sulfur (S), rather than on silicon. We can now expand our list of relevant minerals:

Sedimentary minerals and their chemical composition:

Calcite	$CaCO_3$
Dolomite	$CaMg(CO_3)_2$
Anhydrite	$CaSO_4$

Continental-crust minerals and their chemical composition:

Quartz	SiO_2
Potassium feldspar	$KAlSi_3O_8$
Sodium feldspar	$NaAlSi_3O_8$

Oceanic-crust minerals and their chemical composition:

Olivine	Mg_2SiO_4
Pyroxene	$CaMgSi_2O_6$
Calcium feldspar	$CaAl_2Si_2O_8$

It is finally clear why we drew the wrong conclusion. Nature misled us by mixing sedimentary rocks rich in calcium and magnesium together with the underlying continental crust, which was rich in silicon. The impact energy melted this mixture of very different rocks, which by chance had a *combined* chemical composition close to that of oceanic crust. This chance combination of chemical elements was mixed together in the molten droplets, which were blown right through the atmosphere and launched into space. As those droplets cooled in free fall, outside the atmosphere, before settling back to Earth, they crystallized into olivine, pyroxene, and calcium-rich feldspar. These are the characteristic minerals of ocean crust, and thus we were fooled into thinking the target was in the ocean.

HAD THE IMPACT SITE BEEN SUBDUCTED?

In the early 1980s, the chemical evidence for impact in the ocean seemed so compelling to me that I wasted a great deal of effort in searching for a huge crater on oceanic crust. You would think that a crater 150–200 km in diameter would be easy to find on the ocean floor. Certainly in a well-studied ocean like the Atlantic, crossed by hundreds of oceanographic ship tracks, it would have been found long before—but there was no big crater in the Atlantic. We understood that smaller impacting comets and asteroids would have made their entire crater in the water, leaving no trace on the floor of the ocean, but the KT impactor had an estimated diameter of 10 kilometers—twice the depth of the deep ocean—and surely it would have made a giant crater on the ocean floor.

So maybe the impact was lost in a remote and little-known part of the ocean, like the southernmost Pacific, near Antarctica, where oceanographic ships rarely go. Even so, there should have been sediments recognizable as having been deposited close to the crater, and there were enough deep-sea sediment cores available to show that no such deposits were present in the Pacific. Chemistry said the impact was oceanic, but sediment cores said it was not in any existing ocean. How could we explain that?

Well, Nature had provided us with a ready excuse for not finding the expected oceanic crater. One-fifth of the ocean crust that existed at the time of the Cretaceous-Tertiary boundary has been subducted since then—completely swallowed up into the deep Earth—and if the crater we were seeking had been on that lost crust, it would have been totally destroyed. The 20 percent chance that the crater had been subducted offered us an excuse for not finding it and allowed us to relax and not work as hard as we should have in the search for the impact site.

Sometimes I would think about the huge tsunami that an oceanic impact would produce, and would search the geological literature for signs of the deposits of huge waves around the rim of the Pacific at KT boundary time, but I could not find any. Tsunami deposits would, in the end, be the key to finding the impact site, but it was not to happen for many years, and would be in an unexpected place.

SHOCKED QUARTZ
AND A CONTINENTAL IMPACT

Just as we became comfortable with the evidence that the impact had taken place in an ocean, the picture was suddenly confused with contradictory clues, arguing for a continental hit. Chuck Pillmore's nonmarine KT site in New Mexico had been augmented by many more nonmarine sites in the Rocky Mountain region, from New Mexico and Colorado to Wyoming and Montana, and northward into Sasketchewan and Alberta. A group led by Bruce Bohor, a geologist with the U.S. Geological Survey at Denver, discovered that these sites

A scene from the last day of the Cretaceous in western North America—a *Tyrannosaurus* brings down an *Edmontosaurus*. *(Reconstruction by Dale Russell, painting by Ely Kish)*

The Earth on the last day of the Cretaceous, before the impactor entered the atmosphere. *(Painting by William K. Hartmann)*

The Earth two seconds before the end of the Cretaceous. Scientists are still not sure if the impactor was a comet or an asteroid. What they do know is that the object was at least 10 kilometers wide, and that it entered the Earth's atmosphere at a speed of between 30 and 70 kilometers per second. *(Painting by William K. Hartmann)*

The moment of impact. The first wave of destruction was caused by the huge ejecta curtain, which started to form almost immediately after the impact. *(Painting by William K. Hartmann)*

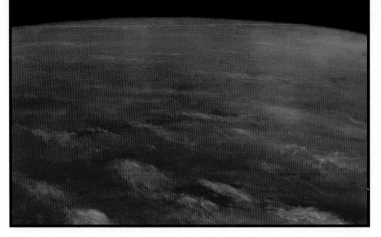

The impact winter. The several waves of destruction caused by the impact produced so much atmospheric debris that the entire Earth was covered by a shroud of dust. This shroud completely blocked sunlight, plunging the surface of the Earth into total darkness for at least several months. It was this "impact winter" that probably caused the extinction of the dinosaurs, and half of all other life forms on the planet. *(Painting by William K. Hartmann)*

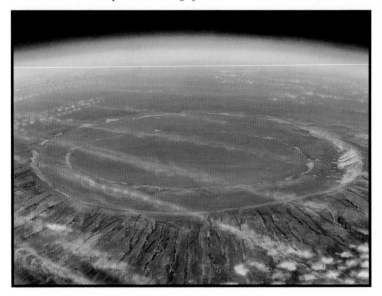

The crater of doom. A view of the immense crater after the sunlight had returned to the Earth. The impactor struck, and the crater was formed, on the north coast of what today is the Yucatán Peninsula. *(Painting by William K. Hartmann)*

A huge tsunami was created by the impactor. Possibly several kilometers high, the tsunami would have destroyed all of the coastlines near ground zero, engulfing dinosaurs living in these areas. *(Paintings by Don Davis,* left, *and Ron Miller,* below*)*

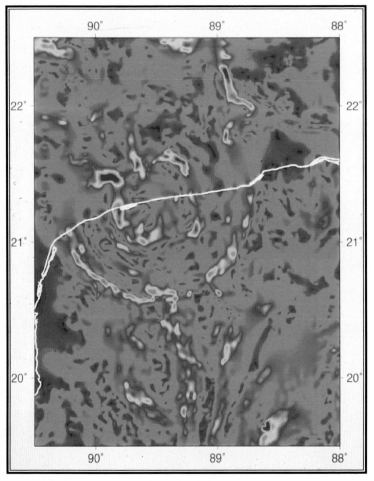

A map of the gravity gradient at Chicxulub.
(Reprinted courtesy of Alan Hildebrand of the Geological Survey of Canada)

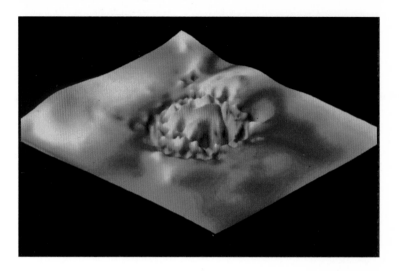

Two perspective plots of the magnetic anomaly field, *above*,
and the gravity field, *below*, at Chicxulub.
(Reprinted courtesy of Mark Pilkington of the Geological Survey of Canada)

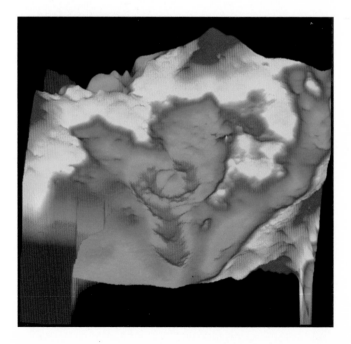

The Shoemaker-Levy team, which discovered the fragmented comet that hit Jupiter in 1994, now consists of *(left to right)* Gene Shoemaker, his wife Carolyn, David Levy, and his wife Wendee. *(Photograph by Jean Mueller)*

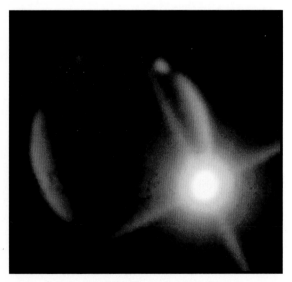

A fragment of the Shoemaker-Levy 9 comet
striking Jupiter, July 1994.
*(Photograph by Peter McGreger,
from Siding Springs Observatory, Australia)*

contain grains of quartz which have been damaged in an unusual way, with multiple sets of planar deformation bands that can be seen under the microscope.[22] This kind of damage was known to occur in quartz grains from bedrock close to proven impact craters and was thought to be due to the shock wave from an impact passing through the surrounding target rock.[23]

Bohor concluded that the quartz grains in the KT boundary had been shocked, and this strongly supported the impact hypothesis, because impact is the only mechanism known to produce shock waves in rocks. Ordinary displacements of the Earth's crust, like those that occur during earthquakes, produce seismic waves that compress and distort rock masses as they pass through them. But like a spring, the rock regains its original shape after the seismic wave has passed. Shock waves are much more intense, and they permanently crush the rock, leaving signs of damage like the planar deformation bands in shocked quartz. Bohor and his Geological Survey colleague Glen Izett did detailed studies of the shocked quartz, making the case for impact shock very convincing.[24]

Other geologists, like Neville Carter and Charles Officer, challenged Bohor and Izett, arguing that the damaged bands in quartz can be produced in volcanic eruptions. They showed photographs of damaged quartz grains from volcanic rocks,[25] but never quite matched the multiple sets of planar deformation bands characteristic of quartz from the KT boundary and from known impact craters. It seemed reasonable that impacts and volcanic explosions should produce different kinds of damage. Volcanic eruptions are not even explosions. They are decompression events, and they do not produce shock waves in rocks. The argument culminated in a showdown at the second Snowbird Conference in 1988, where microscope pictures of volcanic quartz grains were intensely scrutinized to see whether their deformation bands were planar and occurred in multiple sets, as in the shocked quartz from impact craters. When the dust had settled, most of the participants agreed that true shocked quartz could be confidently distinguished from quartz in volcanic rocks, and that true shocked quartz was definitive evidence for an impact, but the agreement was never unanimous.

The shocked quartz gave at least tentative support to the impact hypothesis, but it raised a serious problem. Quartz is the quintessential mineral of the continents and is absent in oceanic crust. If the impact was in an ocean, how could there have been any quartz there to get shocked? The only way out that I could see was for the quartz to have been a deep-ocean sediment, lying on top of the ocean crust at the place of impact. Some of us kept looking for an oceanic site, but other geologists found the shocked quartz to be compelling evidence for a continental hit. They were eventually shown to be right, but the conflicting evidence kept the search for the crater in a state of confusion for several years.

MORE DOUBTS—INDIA
AND THE VOLCANIC SUSPECT

Throughout the 1980s the KT debate was largely polarized between those who thought the KT extinction was the result of impact, and those who attributed it to massive volcanism. The strengths and weaknesses of the two positions were more or less opposite to each other. We in the impact camp had evidence, from the anomalous iridium, spherules, and shocked quartz in the KT boundary layer, for the impact of a comet or an asteroid, but we could not locate the giant crater that would have resulted from the impact. The supporters of volcanism had no strong evidence in the boundary clay to support a giant eruption at KT time, but they could point to a huge volcanic outpouring of roughly the right age in India, called the Deccan Traps. This enormous pile of basalts, covering much of western India, was known to date from roughly the time of the KT boundary.

Dewey McLean at Virginia Tech, reviving an idea of Peter Vogt, argued for a link between the Deccan basalts and the KT extinctions.[26] Dewey suggested that huge amounts of CO_2 had been released by the Deccan volcanism, triggering greenhouse heating that could have caused the extinctions. I countered that the extinction had been much too rapid to result from the eruption of basalts which probably had taken at least a million years, and what's more the Deccan Traps were not very well dated. In Dewey's view, however, the extinctions were not abrupt, but had lasted hundreds of thousands, or perhaps millions of years. Dewey and I had come to completely opposite views of the KT boundary, and our heated exchanges enlivened a few scientific meetings.

But even as the evidence for impact at the KT boundary was building up, so was the evidence that Dewey McLean was right about the age of the Deccan Traps. Vincent Courtillot, a major figure in both science and government in France, began a program of intensive age dating of the Deccan basalts, and the more dates he obtained, the more they homed in on the KT boundary.[27] Vincent and I had worked together in California some years before, and we remained friends, but he was impressed by one part of the evidence and I was impressed by a different part. We also had some exciting confrontations at meetings.

However, the banner of the volcanism supporters was borne most tirelessly by Chuck Officer at Dartmouth. After a distinguished career as an academic and industrial seismologist, Chuck turned his full attention to the KT mystery in the early 1980s. In 1983 and 1985, he and his Dartmouth colleague Charles Drake published two long and detailed critiques of the impact hypothesis, probing every possible weakness in the evidence.[28] Chuck Officer disagreed intensely and often—not only with me, but with almost everyone else who favored impact. Again and again he made us go back and test whether our arguments were really as strong as we thought. Even though it was frustrating not to find the crater for ten years, it was actually a blessing, for an early discovery of the impact site might have short-circuited the intense challenge to each bit of evidence that Chuck Officer compelled us to face.

The discovery of the Yucatán crater makes it hard to continue arguing that the KT extinction was the result of Deccan volcanism. However, as we shall see in chapter 7, it is too soon to say that volcanism plays no role in mass extinctions.

NEMESIS, THE DEATH STAR

Meanwhile, Nature had other intriguing puzzles to complicate the situation. The KT boundary is only one of several known mass extinctions, and in 1984, University of Chicago paleontologists Dave Raup and Jack Sepkoski reviewed the fossil record and suggested that extinctions occur on a regular 26-million-year cycle.[29] Dave Raup sent an advance copy of the paper to Dad, but Dad was sure it had to be wrong. What could possibly cause periodic extinctions on a clocklike timetable? He was pretty sure that at least the KT extinction was caused by impact, and what could be more random than impacts of large asteroids or comets on Earth?

Dad asked Rich Muller to look over the Raup-Sepkoski paper and his own negative response to it, but as Rich analyzed the data closely, he became more and more convinced that Raup and Sepkoski were seeing a real periodicity in extinctions. Dad challenged him to explain how impact-generated extinctions could occur at fixed intervals, and Rich came up with the idea that the Sun might have a distant companion star that comes close to the Sun every 26 million years, somehow triggering a flurry of impacts. A companion star, orbiting the Sun, would be different from all other stars, which move independently of the Sun.

How a companion star could trigger a flurry of impacts remained vague until Rich started exploring the question with astronomers Marc Davis and Piet Hut. As the three of them puzzled over the problem, they realized that although the hypothetical companion star would never come close enough to the Sun to disturb the asteroids in the inner solar system, at its closest passage it could gravitationally alter the orbits of the comets on the very outer fringe of the solar system. This would send some fraction of these comets in close to the Sun and this comet shower would increase the probability of an impact big enough to cause a mass extinction. All the calculations worked out right, and in their paper proposing this mechanism for periodic mass extinctions, Davis, Hut, and Muller proposed the name Nemesis for the very tiny, dim, and inconspicuous companion star of the Sun which might be out there, undiscovered.[30]

I suggested to Rich that if there really was a Nemesis, triggering periodic comet showers, the ages of impact craters on Earth should show the same periodicity, and when we looked closely at the ages in Richard Grieve's crater list, that seemed indeed to be the case.[31] Other possible explanations for extinction periodicity were put forth[32] and there was a furious debate over the validity of the statistical evidence. The story of this episode is well told in Rich's book, *Nemesis, the Death Star*, which is full of insights into how science really works.[33]

Rich began a systematic search for Nemesis, a project very much like looking for a needle in a haystack. He has not found it yet, but someday he may. I suspect that scientists of the future will look back on this episode and be amused, but I'm not sure whether the joke is that a few of us fell for some phony indications of periodicity and dreamed up a crazy story about an imaginary companion star, or that most scientists didn't take it seriously and so the companion star out there, which would change our whole conception of the solar system, has never been found.

STILL MORE DOUBTS—DO IRIDIUM ANOMALIES REALLY MARK IMPACTS?

Even as the KT impact hypothesis was being elaborated into an inference about periodic extinctions and a companion star to the Sun, the initial indication of a KT impact, based on the iridium anomaly, was under attack. Supporters of volcanism seemed vindicated when measurements on gases escaping from the volcano of Kilauea in Hawaii showed the presence of iridium.[34] However, soon there was counterevidence, demonstrating that iridium in volcanic gases had nothing to do with the KT iridium anomaly.

Iridium is one of the six platinum-group elements, all of which are attracted to molten iron and thus are concentrated in the Earth's core and essentially absent at the surface. All of them can be delivered to Earth by impacting asteroids and comets. The young Russian physicist George Bekov was part of a group in Moscow which developed a remarkable analytical technique called laser photoionization, which was suitable for measuring platinum-group elements,[35] and he and Frank Asaro worked together to measure three of them (iridium, ruthenium, and rhodium) in the KT boundary. George and Frank found that in the KT anomaly those elements occur with the same ratios as in meteorites. The ratios among platinum-group elements in volcanic emissions are totally different, because they behave differently in chemical processes that go on within the Earth. The ratios thus gave George and Frank a fingerprint, tying the iridium anomaly to the impact of an extraterrestrial object rather than to a volcanic eruption.

Meanwhile, there was one nagging question through the 1980s which needed to be answered. Many scientists had asked whether the KT iridium anomaly was unusual in the stratigraphic record. "How do you know that there aren't iridium anomalies at many stratigraphic levels, due to some common terrestrial cause like volcanism?" they would ask. It was no use to say that each iridium analysis done by neutron activation, at Berkeley or at one of the few other labs that could do the work, was time consuming and expensive, and that we just couldn't afford to analyze closely spaced samples through hundreds of meters of stratigraphy, looking for stray anomalies.

So my father set out to invent a way of doing lots of iridium analyses quickly and cheaply. It took Dad months of hard work, but finally he was able to combine several clever analytical strategies in a design for a neutron activation machine that could

Sandro Montanari in his mobile field laboratory in the Apennines.

mass-produce iridium analyses. In 1986 Dad's special iridium coincidence counter was ready to take measurements. Now we could hunt systematically through the stratigraphic record to see if iridium anomalies are common or rare.

The obvious rock sequence to search was the Scaglia rossa limestone at Gubbio, where the first KT iridium anomaly had been found. Sandro Montanari did the sampling, collecting hundreds of little pieces of Scaglia limestone, so closely spaced that no significant iridium anomaly could hide between them. Frank and Helen ran limestone samples through the coincidence counter for months, and when they had finished it was clear that there were no other iridium anomalies in that time interval at all comparable with the dinosaur killer.[36] Big impacts are rare. Dad's skill as an inventor had made it possible to tie up a major loose end.[37]

THE BURIED CRATER AT MANSON

There was still another problem in the search for the KT crater. We were not really sure whether we should be looking for a single enormous crater, or whether there might be two or more craters, perhaps smaller in size. The possibility of multiple craters was tied into the idea of comet showers—with or without Nemesis—because a comet shower might yield several impacts close in time to the KT boundary. In 1984 my father organized a meeting of people involved in the question of periodic mass extinctions. At that meeting Rich first pointed out that a comet shower could yield multiple impacts on the Earth over something like a million years, and suggested that such a cluster of hits might explain the gradual extinctions that Bill Clemens and other paleontologists were inferring from the fossil record.[38]

That idea, of course, implied that apparently gradual extinctions really occur as several abrupt steps, closely spaced in time. University of Colorado paleontologist Erle Kauffman has led the effort to determine whether this is so—a difficult task that is right at the limit of resolution of the fossil record, and is probably different for different mass extinctions.

It also implied that crater ages should be closely clustered in time and that layers of impact ejecta should be bunched in the stratigraphic record. This became a particular interest of Gene Shoemaker and Sandro Montanari. The work they have done with various colleagues has made it pretty clear that there was a cluster of impacts during a time of increased but noncatastrophic extinction near the Eocene-Oligocene boundary, about 34 million years ago.[39]

The idea of comet showers and multiple impacts also suggested a possible solution to the old problem that the KT spherules indicated an impact in the ocean, while the shocked quartz argued for a hit on a continent. Maybe there had been two hits—one oceanic and one continental. Indeed, the KT boundary in the western United States has two layers—a lower one with spherules and an upper one with shocked quartz. The two layers are in contact with each other, but cleanly separated.[40] They really look like they are due to separate impacts. And moreover, there was a nearby crater that looked like it could be the source of the shocked quartz of the upper layer.

Under the glacial drift of the farmlands of central Iowa, near the town of Manson, the continental crust of North America bears the scar of a large impact. The Manson crater is 35 km across—not big enough to cause a mass extinction, because there are more craters this size than there are mass extinctions, but substantial nevertheless.[41] Preliminary age dates showed that Manson was of about the same age as the KT boundary, and the bedrock is rich in quartz. At last, it seemed, Nature's trick had been figured out. There were two KT hits. The continental hit was at Manson, and the oceanic one was probably on crust which had been subducted and would never be found. It was satisfying to have finally understood Nature's ploy, but the satisfaction was premature. Nature was about to have another laugh at our expense.

Gene Shoemaker, David Roddy, Ray Anderson, and Jack Hartung organized a drilling program at Manson, and they recovered spectacular impact-shattered rocks, full of shocked quartz.[42] When the shocked rocks were dated, they gave an age of 74 million years, decidedly older than the KT boundary at 65 million years. The new date suggested where in the rock record to look for the debris, and sure enough, in South Dakota Glen Izett found the ejecta from Manson, well below the KT boundary.[43]

So Manson had been just another red herring in the KT mystery. But by the time Manson was removed from the list of suspects, we were on the right track at last.

Dad never found out about it. He passed away in 1988. For ten years he had been at the center of some of the most exciting research on Earth history. He had delighted in the effort to get past all the tricks and stumbling blocks Nature had placed in the way of finding the site of the impact. Dad would have loved the discovery of the Crater of Doom.

The Crater of Doom

Throughout the decade of the 1980s, more and more evidence was discovered that supported the impact theory for the KT extinction, but the impact site remained frustratingly elusive.

In a good mystery story where the crime is concealed almost perfectly, there is usually a red herring to confuse the detectives. In our case the red herring was the misleading evidence, described in the previous chapter, that pointed to impact in the ocean. However, in a good mystery, there is one tiny flaw in the concealment. Eventually the detective finds the flaw, the rest of the disguise crumbles away, and the culprit is finally exposed. That's the way it was with the search for the KT crater. The disguise was almost perfect, but the flaw was the tsunami.

A giant oceanic impact would have generated a truly enormous tsunami, capable of eroding the floor of the deep sea at depths no other waves ever reach. When the tsunami reached the continental margin, it would build up into a towering wave, perhaps a kilometer high, that would crash down near the shore. Coastal forests would be destroyed and coastal sand would be shaken loose and slide down into deep water as the giant, fluidized, submarine landslides that geologists call turbidity flows. Turbidity flows deposit sand beds called turbidites. If we could find an exposure of marine sediments near the edge of an ocean, with a turbidite right at the boundary, it would point to that ocean as the site of the impact.

But as we now know, the impact was not in the ocean. It was on the continental crust of the Yucatán, above or just slightly below sea level, where no giant, deep-water tsunami should have been generated. If the concealment of the crime had been perfect, there would have been no tsunami deposit anywhere, and we would have gone on looking for it indefinitely, and in vain.

However, there was a tiny flaw in the concealment. The impact was on the continent, but it was close to the ocean. It was close enough that a tsunami was generated in the adjacent ocean anyway—perhaps by debris from the crater falling into the deep water nearby, or from seismic waves or submarine landslides triggered by the impact. The exact mechanism is not yet clear, but immediately after the comet hit the Yucatán, the tsunami sped away from the impact site. It left evidence of its passage in the form of a torn-up sea bottom covered by sedimentary debris—the evidence we were seeking. We had been fooled for years, but we were about to stumble onto the flaw in the nearly perfect concealment.

HAITI AND TEXAS

Florentin Maurrasse is a Haitian-American geologist at Florida International University, and we have been friends since the 1970s, when we were both researchers at Columbia University's Lamont-Doherty Geological Observatory. Many years ago, Florentin discovered a deep-marine KT boundary site near the town of Beloc, on the Southern Peninsula of Haiti. There is a coarse, sandy bed right at the boundary in the Beloc outcrop, but unfortunately Florentin found it before many people cared about the KT boundary, and before anyone knew the critical questions to ask. He published a paper on Beloc in 1980[1]—the same year as the original iridium discovery papers— and then, learning of our work, he sent samples to Berkeley. Frank and Helen found an iridium anomaly, so Beloc became one of the early confirmation sites.[2] But this was too early, before we realized that we should be looking for a turbidite. At the time the sandy boundary bed at Beloc did not seem particularly important. As the search for more KT localities continued, Beloc was just another deep-water KT iridium site on a growing list.

It was remote, few geologists had been there, and no one realized that Beloc held evidence for the tsunami. The main clue would come from somewhere else.

The Brazos is one of the many rivers that flow south across Texas and empty into the Gulf of Mexico. The coastal-plain sediments slope very gently southward, so the Brazos passes stratigraphically up through younger and younger beds on its way to the Gulf. There are not many outcrops in these soft sediments, but between Waco and College Station the river tumbles over some low rapids formed by a hard sandy bed. In the early 1980s this area attracted the attention of Thor Hansen, a paleontologist at the University of Texas. Thor made detailed fossil collections which showed that the sandy bed was right at the KT boundary and he recognized that it was different from the fine-grained marine sediments above and below.[3] Ted Bunch and Rosalie Maddocks sent samples to Frank and Helen, who found the expected iridium anomaly. Just as in the case of Beloc, it was too early to appreciate the importance of the sandy bed.

I believe the first person to sense the possible significance of the Brazos sandy bed was Jan Smit. Jan has studied more KT boundaries around the world than anyone else, and when he first went to the Brazos River in the early 1980s, he recognized that the sandy bed was something unusual. In a 1985 paper with Ton Romein,[4] Jan included this comment about the Brazos outcrop: "This may be the first evidence of impact (?tsunami)-triggered sediment."

Even though tsunami waves have been carefully studied because of the hazard they pose to coastal populations, we know almost nothing about their deposits, for few if any have been recognized in the stratigraphic record. Even an experienced sedimentologist like Jan could not know what features might indicate deposition by a huge tsunami. Finally the Brazos bed came to the attention of Jody Bourgeois, a sedimentologist at the University of Washington. Jody had been a student at Columbia when Florentin and I were there, and she was particularly interested in the deposits of giant storms. Probably more than any other sedimentologist, she knew what little there was to know about tsunami deposits. Jody gathered a team to study the Brazos locality in detail, and it was clear from their work that only a really enormous tsunami could explain the detailed characteristics of the Brazos sand bed.[5]

It is ironic to see how slow we were to appreciate the significance of the sandy beds at the Brazos River and Beloc. In retrospect, Maurrasse or Hansen or Smit, or we at Berkeley with the samples for iridium analysis—any of us could have figured out that the impact site was in the Gulf of Mexico-Caribbean region several years earlier than actually happened. But we did not, because the real clues were inconspicuous among the masses of data that hundreds of scientists were publishing. By the late 1980s we knew of more than 100 KT sites with iridium and other interesting evidence of all kinds, and the Brazos River did not stand out as particularly significant until Jody Bourgeois and her colleagues showed that it contained a tsunami deposit exactly at the KT boundary.

At that point, what was needed was someone absolutely focused on finding the source of the Brazos tsunami—someone who was persuaded that the Brazos sand bed was the fundamental clue and who would not rest until the culprit was tracked down. That's what I should have been doing, but I was more interested in the well-known candidate sites like the vast outpouring of Deccan lavas in India and the Manson Crater in Iowa. And always I suspected that the KT impact site had really been on oceanic crust that had been subducted. It was Alan Hildebrand who would be that relentless detective.

ALAN HILDEBRAND'S SEARCH
FOR THE CRATER

Alan is a Canadian who had come to the U.S. in the early 1980s to study with Bill Boynton at the University of Arizona. The critical task for a new graduate student is to pick a topic for a Ph.D. thesis which is sufficiently challenging and significant, but not so difficult as to be impossible. Alan focused on the KT boundary from the beginning of his graduate work. Feeling his way toward the heart of the problem, he looked first at the possibility of impact-generated volcanism, and then found some more of the misleading evidence for an oceanic impact.[6]

By 1988, Alan had decided that the Brazos River tsunami bed was the key to finding the crater. He knew that the tsunami could only have come from south of Texas, because that was the direction toward deep water 65 million years ago, just as it is now. He reasoned that the impact site could not have been too distant from Texas, because the Gulf of Mexico is an enclosed body of water, protected from any tsunami that came from far away. Accepting the prevalent view that the impact was on ocean crust, Alan focused his attention on the Gulf of Mexico and the Caribbean.

In a dogged search, Alan returned to the Brazos River again and again, trying to extract every obscure hint and every last shred of evidence from the tsunami deposit, and he combed the published literature and maps of the Gulf and the Caribbean for any sign of possible impact debris, or for any large circular structure which might be an impact crater. He found a vaguely rounded set of features on maps of the floor of the Caribbean north of Colombia and learned of a pattern of circular gravity anomalies at the north coast of the Yucatán Peninsula. The Yucatán candidate looked really promising, even if it was on continental crust.

At meetings in 1990, Alan gave talks about what he was doing, and he started to get other people interested in the Gulf of México and the Caribbean. For some reason, I had never been particularly impressed with what I had heard about the Brazos River, but one day in early 1990 I had a new idea for a way to look for evidence of a tsunami—not by looking for the sedimentary deposits of the tsunami, but by looking for a gap in the sedimentary record due to tsunami *erosion*. I reasoned that an impact in an ocean would send tsunamis crashing into all the surrounding shorelines, eroding the continental-margin sediment. After the event was over, deposition would resume, and

Philippe Claeys (*left*) and Alan Hildebrand at the KT boundary out-crop on a side valley of the Brazos River in Texas, where the first tsu-nami deposit was recognized.

the result would be an unconformity—a gap in the sedimentary record—with the upper part of the Cretaceous missing, but the very basal Tertiary present. Even if the impact site had been on oceanic crust that had been subducted, tsunami erosion of the surrounding continental margins might reveal where the crater had been.

I scanned through the records of the hundreds of sediment cores taken by the Ocean Drilling Project, and there was only one place in the world with that kind of a gap in the record—it was the Gulf of Mexico. As soon as I could, I went to the core archives at Lamont-Doherty to study and sample the Gulf of Mexico cores from Drilling Leg 77. Just above the gap where the upper half of the Cretaceous was missing there was a strange bed of sand—with ripples that showed strong currents in this normally quiet, deep-water environment, and full of clay specks that just might be altered glass. Could this be a deposit of impact-melted glass particles stirred around by an impact tsunami? Suddenly I started taking Alan Hildebrand's ideas very seriously.

Almost nothing had been published in the literature about the circular pattern of gravity anomalies in the Yucatán which suggested a buried crater, and Alan had to do real detective work in order to find out about them. At last he tracked down the people who knew about the Yucatán structure, and thus he was the first of the KT researchers to meet Antonio Camargo and Glen Penfield. Finally, in 1991, a paper was published by Hildebrand, Penfield, Kring, Pilkington, Camargo, Jacobsen, and Boynton, entitled "Chicxulub crater: a possible Cretaceous/ Tertiary boundary impact crater on the Yucatán Peninsula, Mexico."[7]

It was a bombshell. The Crater of Doom was found at last! The clue had been the tsunami, generated even though the impact had been on continental crust. Nature had buried the crater and it was completely invisible at the surface, but the tsunami had spread the evidence of nearby impact to an outcrop in Texas. Thor Hansen's fossil age, Jan Smit's hunch, Jody Bourgeois's detailed study, and Alan Hildebrand's relentless search had come to fruition. We learned to spell Chicxulub, found that it was a Mayan word pronounced "Cheek-shoe-lube," and began to hear the remarkable story that Glen Penfield and Antonio Camargo had known for ten years.

ANTONIO CAMARGO AND GLEN PENFIELD

In the opening scenes of *The Treasure of the Sierra Madre*, Humphrey Bogart is an American worker in the depression-era oilfields at Tampico, the Mexican petroleum capital on the Gulf Coast. Soon after the time depicted in the film, in 1938, the foreign companies developing the oil were thrown out of the country by President Lázaro Cárdenas. Proud, independent Mexico decided to go it alone, Petróleos Mexicanos (PEMEX) developed into a giant national oil company, and for 50 years

geologists outside Mexico knew little about what was being discovered there.

Mexican geologists and geophysicists explored their own
country for petroleum and discovered huge oil fields. One place
they were not successful was on the flat northern coastal plain of
the Yucatán, although it looked promising at first. In this fea-
tureless landscape, the first step in finding oil is to make a gravity survey, mapping tiny variations in the pull of gravity which
reflect variations in rock density at depth, which in turn may
reveal buried structures that might contain oil. The initial gravity survey of the Yucatán[8] revealed an enormous circular structure, buried below the surface and centered at Puerto Chicxulub, on the north coast near Mérida.

I imagine that the PEMEX geologists were extremely excited
over the oil potential of this huge gravity feature. Yet when they
drilled the structure in 1952, their optimism must have turned to
disappointment. After penetrating about a kilometer of Tertiary
sediments, the drill began to bring up pieces of hard, dense,
crystalline rock—very different from the porous sedimentary
rocks in which oil is found. Chemical analyses gave a composition similar to andesite, the common volcanic rock which is
spread over much of western North America and forms the volcanoes that overlook Mexico City. The PEMEX geologists concluded that they had discovered a buried volcano. One does not
find oil in volcanoes, and after several dry holes, the Yucatán
project was terminated. We now know that it was not a volcano,
but we cannot criticize the PEMEX geologists, for in 1950 probably no more than a half dozen people in the world could have
recognized that the crystalline rocks were not volcanic andesite,
but were impact-melt rocks.

PEMEX scientists figured out the right explanation before
anyone else did, however. Antonio Camargo Zanoguera, a geophysicist born in 1940 inside the Chicxulub ring, and Glen Penfield, an American geophysicist consulting for PEMEX, under-
took a detailed restudy of the northern Yucatán in the 1970s.
The Chicxulub structure had none of the characteristics of a

volcano except for the andesite, and trying to explain all its strange features, they began to wonder if it could be an impact crater. They studied all the publications they could find on impact structures, and everything fit, except that Chicxulub was very much larger than any known impact crater on Earth.

Academic geologists are expected to report the results of their research in the scientific literature, but those who work for oil companies publish less often, because much of the information they deal with is confidential. Penfield and Camargo gave only one brief talk, in 1981, accompanied by an abstract in the program of a meeting.[9] What an irony! The previous year we had published the evidence for a giant KT impact, but it took ten years to put the two together.

In retrospect, I think the long delay in connecting Chicxulub with the KT boundary was a good thing. Hundreds of careful investigations were done on the problem during those ten years before the crater was identified. As a result we know much more about the KT boundary event than we would have if an early discovery of the crater had made it a problem of less pressing interest. And yet, during all those years of our fruitless search, a magnificent study of Chicxulub lay hidden away in the files of PEMEX. Finally, after Alan Hildebrand brought the Yucatán crater to the attention of the KT boundary scientists, Camargo and Penfield at last began to talk about their work. Much later, at the third Snowbird Conference in Houston in 1994, Antonio Camargo presented their study in full, impressing the audience with the level of detail and sophistication with which he and Glen Penfield had understood the KT crater, 13 years earlier.

THE TSUNAMI BED AT ARROYO EL MIMBRAL

The recognition of the Chicxulub crater changed the direction of KT research. Many of us wanted to analyze the impact melt rocks, but the crater is deeply buried and we could not just go and collect samples. Cores from the old PEMEX wells were suddenly in great demand. Unfortunately there was a distressing report that all the cores had been destroyed in a warehouse fire. Glen Penfield thought there might still be some bits of lost core lying around the old well sites in the Yucatán, and later I saw some very amusing documentary film footage of Glen digging through a heap of pig manure where the villagers said the drill rig had been, 30 years earlier. Unfortunately, Glen found no cores in the pig manure. "Geology is not as elegant a science as physics," I told Rich Muller.

For the foreseeable future, the deeply buried crater was out of reach. What could we do in the meantime? By good fortune, Jan Smit was in Berkeley for a few months, starting in December of 1990, just as the Chicxulub business was coming to a head. Jan, Sandro Montanari, and I asked ourselves what we could do on a low budget to test whether Chicxulub was really the KT impact site. The biggest question was the age of the crater. Was it precisely the same age as the KT boundary? Or was it older or younger, and therefore unrelated to the extinctions? This would be the critical test.

The crater was buried and inaccessible, so we decided to search for the closest place to Chicxulub where KT-age sediments might outcrop at the surface, so that field geologists like us could get at them. It looked like the best place would be northeastern Mexico, where sediments deposited on the floor of the Gulf of Mexico in the late Cretaceous and early Tertiary were later uplifted and are now exposed in a semiarid desert. It was hard to find published papers about the geology there. Surely PEMEX geologists knew all about the area, but their studies must have been in company reports, not in the international literature. We combed the Earth Sciences Library at Berkeley and the results were disappointing. Almost the only thing we could find was a book by the American geologist John M. Muir (not the famous naturalist), dating from 1936, back in Humphrey Bogart days, before the Mexican petroleum industry was nationalized.[10] It's not often in modern science that a 50-year-old book is the key to discovery, but we were intrigued by Muir's description of what sounded like a sandy bed at what might be the KT boundary near Ciudad Victoria. We decided to go and look for KT outcrops all over this part of Mexico, and as we planned our trip, we were helped a lot by suggestions from José Longoria and Marta Gamper, a husband-and-wife paleontologist team who had long studied the microfossils in northeastern Mexico and knew the area well.[11]

In February of 1991, Jan, Sandro, and I set out, along with Milly and an English postdoctoral researcher named Nicola Swinburne, to look for KT boundary outcrops in northeastern Mexico. For days we searched in vain through dramatic mountains and deserts, where the memories of Pancho Villa and the Mexican Revolution live on in the irresistible music of the corridos. Checking off candidate sites where we could not identify the KT boundary, we worked our way south and our discouragement grew. Years earlier, on separate trips in this part of Mexico, Jan and I had seen a couple of rather ordinary-looking KT boundary sites, not suggestive of tsunami deposition. Now we were finding nothing at all. If the KT boundary here, only a few hundred kilometers from Chicxulub, was quiet and undisturbed, Chicxulub could not be the KT impact site, and we would be back to zero.

Muir's locality was the farthest south on our list of candidates, and car troubles almost kept us from getting there at all. On our last afternoon we searched for it, far up a dry riverbed named Arroyo el Mimbral. Wherever a patch of mudstone bedrock poked up through the gravel, Jan would study a sample with his hand lens and report the age of the forams—closer and closer to the base of the Tertiary, he said. The jeep track was rough and the beds dipped gently, so we worked our way down through the stratigraphy very slowly, gradually approaching the KT boundary as the Sun began to sink toward the horizon.

Finally the low walls of the arroyo rose into a higher bluff, with the biggest rock exposure we had seen for many kilometers. We hurried toward it with growing excitement and came to the most amazing outcrop I have seen in 30 years as a geologist. We scrambled over the rocks, shouting out one discovery after another as the light faded. "Look at the current bedding in this sand!" "Hey—this bed is packed with spherules!" "What's all this fossil wood doing in these deep-water sediments?!" Reluctantly we started back when we could no longer see anything at all. Arroyo el Mimbral confirmed a well-known law of geology—the best outcrop of a field season is always found on the last day, at the most remote location, just as it's getting dark.

Back in Ciudad Victoria we held a council of war. I fought a struggle with my conscience and made what now seems like the mistake of letting responsibilities interfere with a great scientific opportunity. Milly, Nicola, and I decided to return to Berkeley, but Jan and Sandro changed their plans, bought machetes and cooking pots, and went to camp at Arroyo el Mimbral. Over the next week they studied, measured, drew, sampled, and photographed the outcrop, and the story came into focus.

The site of Arroyo el Mimbral had been way out in the Gulf of Mexico in the late Cretaceous and the early Tertiary. In these deep-water conditions there were no currents capable of transporting sand. Only marl—a mixture of fine-grained clay and calcium carbonate—could be deposited in this setting. The tiny clay particles drift for long distances in the ocean, where they are mixed in with the calcium-carbonate microfossil shells of forams which settle to the bottom after the forams die. The forams made it possible to identify the precise level of the KT boundary, and it was marked by a boundary bed a hundred times thicker than at most other KT localities. At the moment corresponding to the boundary, quiet deposition was interrupted by the emplacement of a three-meter bed of sand—a sediment foreign to the tranquil regime of the sea bottom. Jan and Sandro found three different units, or subdivisions, in the sand bed, and there was no indication of the passage of any significant time between the three units. As far as they could tell, all three units could have been deposited within a few days. Here is what they saw in each unit, and how they interpreted it:

(1) First, at the bottom, is a meter of local deep-water sediment ripped up from the seabed and mixed together with impact spherules and chunks of limestone which we now suspect are ejected fragments of the Yucatán platform. This first unit

Jan Smit points to the base of the KT boundary outcrop at Arroyo el Mimbral in northeast Mexico, where passage of the tsunami wave scoured the bottom of the Gulf of Mexico just after the impact debris had fallen.

appears to reflect the passage of the tsunami, sweeping out from the impact site, violently disrupting the quiet sea bottom while solid and liquid ejecta from Chicxulub rained down through the atmosphere.

(2) Above the tsunami deposit is a complex, two-meter bed of sand with a source quite different from what lies below. The

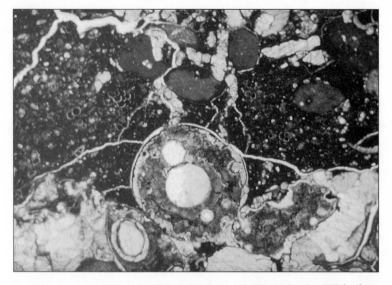

This microscope photograph of the lowest interval in the KT bed at Arroyo el Mimbral includes a round spherule about one millimeter across with bubbles in it. The spherule was originally impact melt but has been altered to clay. The spherule has been pushed into a fragment of seafloor mud in which foram shells are visible. The smaller, grey, angular objects are limestone fragments, probably blasted out of the Yucatán impact site.

sand was derived from what was then the coastline of Mexico, and appears to have been violently shaken up and fluidized by mixing with seawater when the tsunami crashed into the coast. The fluidized sand swept down the steep continental margin in a swift submarine turbidity current, losing energy and settling out as a turbidite bed on the flat sea floor. As a witness to its coastal origin, the sand bed contains layers felted with petrified wood, which is obviously foreign to the deep-water setting and apparently represents the destruction of the coastal forests of Mexico by the crashing tsunami wave.

(3) Finally, at the top, there are alternating beds of rippled sand and fine clay, which probably represent several passes of large waves called seiches, as the broken remains of the tsunami sloshed back and forth in the enclosed Gulf of Mexico. Frank

Asaro found the iridium anomaly just at the top of the seiche deposits. The iridium must occur in tiny particles derived from the vaporized impactor, and these tiny particles could not settle out until the water was completely calm again.

Above this complicated and very informative boundary bed, the deposition of quiet-water marls resumed as if nothing had happened—except that most of the species of forams which had flourished in the surface water of the ocean were now extinct. It would be hard to imagine a clearer testimonial to the KT boundary impact. The ejecta from Chicxulub was all mixed up in a tsunami deposit precisely at the stratigraphic level of the mass extinction.

GLASS!

Sandro and Jan returned to Berkeley with lots of samples, including a huge slab of sandstone that was packed with fossil wood from the destroyed coastal forests. Between the Mimbral collections and the samples I had taken from the Leg 77 cores, we had lots of material to study. The most important thing to look for was glass. When rock is melted—either slowly by igneous heat from the Earth's interior, or suddenly by impact—and it cools rapidly, there is no time for crystals to grow. The result is a glass, with the atoms disorganized instead of arranged in a regular crystal lattice. Glasses preserve the original chemical composition of the melt, but they are unstable and easily alter to clay, so geologically old glass is rare.

For years people had searched for impact glass in KT spherules but had found only alteration products, where the original chemistry had been destroyed. But the hope of finding glass remained alive, and finally in late 1990 and early 1991, four different groups discovered glass in the spherules from Florentin Maurrasse's Beloc site in Haiti.[12] Haraldur Sigurdsson is an Icelandic volcanologist at the University of Rhode Island with extensive knowledge of volcanic glass; Glen Izett of the U.S. Geological Survey had studied volcanic glass and KT shocked quartz;

and John Lyons and Chuck Officer from Dartmouth studied the Beloc site from the anti-impact point of view. Each of them, and Florentin as well, recovered glass from the Beloc spherules at just about the same time. All but the Dartmouth group interpreted the spherules as tektites—droplets of impact-melt glass that had been launched into space beyond the atmosphere and had fallen back to Earth.[13]

There was much excitement as the chemical and isotopic composition of the Beloc microtektites was measured, for it would be a direct reflection of the composition of the target. As far as we knew at that time, no melt-rock samples from the Mexican wells at Chicxulub had survived the PEMEX warehouse fire, so the Beloc glass would be the only clue to what was down there.

Some of the first results came from analyses by Haraldur Sigurdsson and a group of French geochemists,[14] who found that the tektites were mostly made of black glass with a chemistry showing that it was derived from the continental-crust rocks that formed the basement of the Yucatán. Within the black glass there were streaks of yellow glass rich in calcium which came from impact melting of the calcium-rich sediments—limestone, dolomite, and anhydrite—that had been deposited as a thick layer on top of the basement.

At last we could understand how we had been misled for so long. It was an amazing coincidence. Nature had used a mixed target with continental crust and continental sediments which combined to give the chemistry of pyroxene and calcium feldspar—the essential minerals of oceanic crust. A misinterpretation that had confused things for almost ten years was cleared up at last.

The second important implication of Haraldur's work came from the simple fact that the colors in the glass were streaky, which showed that it had not stayed melted long enough to homogenize. This alone argued for an impact event, because impact melts quickly freeze, whereas volcanic melts stay liquid for a long time and are usually homogeneous.

Our group at Berkeley had none of the exciting Beloc glass to work on, but we had the new material from Mimbral. Our samples were packed with spherules, but every one we looked at was altered. We were impressed with the tiny bubbles that occurred in most of the spherules and could still be seen, despite the alteration. We suspected that they were due to the carbon dioxide gas that would be given off when limestone and dolomite, the surface layers in the target rock at Chicxulub, are shocked by an impact.

But we didn't see any glass. Maybe someone else would have better luck. So we sent samples to Alan Hildebrand, who had surely earned the privilege of studying the spherules at Mimbral—the closest known outcrop to his Chicxulub Crater.

Just at that time we were getting to know Stan Margolis, Professor of Geology at the University of California at Davis, his wife and technician, Karen, and his Belgian graduate student, Philippe Claeys. Stan had extensive experience in identifying and analyzing microtektites. He was working at the time with Philippe on microtektites from a young impact in the Pacific discovered by Frank Kyte,[15] and if anyone could find glass in the Mimbral spherules, it would be Stan and Philippe. We gave samples to them as well.

One memorable day in May of 1991, Alan telephoned with the wonderful news that he had found bits of preserved glass in the Mimbral samples. I had no sooner put the phone down than it rang again, and it was Philippe, reporting that he and Stan also had found Mimbral glass. Soon Miriam Kastner, a geochemist at the University of California at San Diego, extracted tiny bits of glass from our Leg 77 samples as well. Jan and Sandro and I should have been more persistent and found the glass ourselves, but I'm glad we didn't, because the friendship and collaboration that grew up with Stan and Philippe was extremely rewarding. That summer we were constantly driving back and forth between Davis and Berkeley as we worked on the analysis of the Mimbral glass. When Stan died suddenly and tragically of cancer in the fall of 1992, Philippe completed his Ph.D. thesis with me and came to Berkeley as a postdoctoral researcher.

While we were studying the Mimbral glass with Stan and Philippe, other people in other labs were analyzing the Beloc glass. It was a time of the highest excitement and colleagiality, as direct chemical clues to the impact unfolded before us. The only thing missing was the old PEMEX cores from the Chicxulub crater. Those lost samples of the melt rock at the impact site would have made all kinds of chemical and isotopic tests possible, allowing us to determine for sure whether the Beloc and Mimbral glass, and all the other KT ejecta, really came from Chicxulub. The film of Glen Penfield searching in vain for melt-rock cores in the pile of pig manure was amusing, but it was also sad and frustrating. The vital evidence lay only one mile away from Puerto Chicxulub, but it was one mile straight down and completely inaccessible.[16]

And then, toward the end of 1991, splendid news came out of Mexico, from José Manuel Grajales, a geologist with IMP, the Mexican Petroleum Institute—the research arm of PEMEX. Manuel had been searching through the sample archives of IMP for the Chicxulub cores, and after some real detective work, he had at last tracked them down! The cores had not been lost or destroyed. They had been carefully curated and stored, but so long before that they were not easy to find.

Some of the cores contained rock which had obviously cooled from a melt, and these first melt-rock samples seemed like the most precious rocks in the world—as rare and informative as the lunar samples. Yet before long Manuel and other Mexican geologists were able to find lots of additional Chicxulub cores. Full testing of the Yucatán-Beloc-Mimbral-KT link was possible at last.

THE SMOKING GUN

About this time, the newspapers began to refer to Chicxulub as the smoking gun in the KT extinction mystery, and indeed the melt-rock samples from the PEMEX cores were the vital clue. Manuel Grajales and his colleague at IMP, Ernesto Cedillo-Pardo, began a study of the melt rock, and scientists in Mexico generously allowed a number of other researchers to work on the precious samples. From laboratories using a whole range of specialized techniques the results began to come in. Two groups—one at Berkeley and Stanford and the other headed by Buck Sharpton of the Lunar and Planetary Institute in Houston and Brent Dalrymple of the U.S. Geological Survey—reported radiometric age determinations which showed that the Chicxulub melt rocks were of KT age.[17]

An even better confirmation of the age of the crater came from chemical studies which showed the same isotopic peculiarities in the melt rock cores from the crater and in tektite glass from Haiti and Mimbral. It became harder and harder to doubt that the KT tektites came from Chicxulub. Since the tektites lie at precisely the stratigraphic horizon of the foram extinction, the crater must have formed at just that time. Behind this conclusion lay a human drama, for some of the key work that confirmed the impact origin of the glass and the isotopic link between Chicxulub and the KT tektites was done at Dartmouth by Joel Blum and Page Chamberlain.[18] Dartmouth had long been identified with the anti-impact viewpoint because Chuck Officer and Charles Drake were based there. Now the two interpretations of the KT event were in open conflict at Dartmouth, culminating in the "Dartmouth Dead Dino Debate" of 1993, featuring Chuck Officer vs. Joel Blum. That's a shootout I wish I could have heard!

CHICXULUB AND MIMBRAL UNDER FIRE

Naively, Jan and Sandro and I imagined that the discovery of the largest impact crater on Earth and the identification of impact glass in a tsunami bed precisely at the KT boundary only a few hundred kilometers away would wrap up the story for good. How foolish of us! Science does not work this way. Every significant conclusion is challenged as severely as possible, and every bit of logic and interpretation is put to the test. This is what happened with Chicxulub and Arroyo el Mimbral, and the challenges and counterchallenges moved so quickly that sometimes it was hard to stay abreast of what was happening.

By the time our papers on Arroyo el Mimbral and Leg 77 were published,[19] opponents of impact were already taking their best shots. Chuck Officer and his colleagues challenged all the evidence for impact in or near the Caribbean, and specifically questioned the impact origin of the Chicxulub melt rock.[20]

Officer was reinforced by foram specialist Gerta Keller at Princeton, long a skeptic, who has emerged as the most energetic opponent of the Mexican evidence for the KT impact. With her colleagues Wolfgang Stinnesbeck and Thierry Adatte, Gerta quickly got to Arroyo el Mimbral, restudied the outcrop, and challenged our interpretation in almost every regard.[21] Much of the ensuing debate hinged on subtleties in the record of foraminifera above and below the three-meter bed of sand. Obviously there would be a major debate over these conflicting views, and we needed more information to be sure. It was clearly necessary to go back to Mexico and look for more KT boundary outcrops, and we needed little further encouragement. Sandro was no longer in Berkeley—he had returned to Italy to start the Geological Observatory of Coldigioco—the first private research and teaching institute for geology in Italy, and soon to become a major center for impact studies. Jan and I, with Milly and Nicola, headed for Mexico.

THE PAVING STONE AND THE BUBBLES

On our previous trip, in February of 1991, we had been feeling our way into an unfamiliar region, not sure where the good outcrops might be located. We had had a succession of disappointments until the last afternoon, when we finally found the KT boundary at Arroyo el Mimbral. Our hope when we returned in January of 1992 was that we might be lucky enough to find one more boundary outcrop. But this trip was different because we knew what to look for, and we were with people who knew the geology very well.[22]

Manuel Grajales came from Mexico City to join us, and he had arranged to work with PEMEX field geologists from this part of Mexico. First we met up with Mauricio Guzmán and Manuel Zambrano, from the office in Tampico. They had not worked on the KT interval themselves, but they had gone over the PEMEX field maps and picked a good place to look. Their target was La Lajilla, a village next to a low dam which impounds a large shallow reservoir. In the morning we drove through heavy rain toward La Lajilla, along a gravel road laid down across a sea of mud. Mauricio and Manuel drove ahead in their jeep, and we followed in our van. The mud had been deposited in the still water on the floor of the Gulf of Mexico, and much later it had been slowly lifted up above sea level. We were picturing some very messy going if the gravel road ended short of the village. Suddenly Manuel Grajales, who had been studying the map, asked a question: "Lajilla is a diminutive of the word *laja*," he said. "Do any of you know what that means in Spanish?" None of us did. Manuel translated: "It means paving stone!"

We looked out at the morasse of mud, thought about what "paving stone" might indicate, and began to feel optimistic. Sure enough, as we rounded a bend just before the village, there it was—a thick bed of sandstone sloping gently up out of the ancient mud of the Gulf of Mexico. It was the paving stone of La Lajilla—the solid ground where the village had been built. The sloping sandstone bed was cut through by an arroyo, and there the dam had been built, anchored to the sandstone outcrops at either end. Indeed, as Manuel had guessed, the sandstone *laja* was the KT boundary bed.

It was another splendid outcrop, carrying the spherules which now seemed like old friends. Jan was especially interested in the current bedding in the sands which had slumped down off the devastated ancient coast of Mexico. The current bedding told a story of repeated reversals in the flow of rushing bottom water stirred up by the tsunami in the Gulf of Mexico. Mauricio Guzmán and Manuel Zambrano had chosen their target well, and La Lajilla has become one of the most informative of the KT boundary outcrops.

A couple of days later and a hundred kilometers to the north, Manuel Grajales arranged for us to join another team of PEMEX geologists, from the office at Reynosa, across the Rio Grande from Texas. In the plaza of a town called General Terán we met Ricardo Martinez, Pedro Romero, and Eduardo Ruiz—skilled field geologists whose job is to map and study the surface outcrops of northeastern Mexico for information that will help in the search for oil in the subsurface. "What exactly are you trying to find?" they asked us.

"It's this peculiar bed at the KT boundary," I explained, holding out a sample from La Lajilla. "It has these little spherules in it. Look—with a hand lens you can see tiny gas bubbles in them. We suspect they're melt droplets ejected from the impact crater at Chicxulub."

Ricardo, Pedro, and Eduardo looked at each other in a strange way, and then Pedro walked over to their jeep. He returned with a big chunk of rock, pointed to a spherule full of tiny bubbles, and asked with a grin, "Is this what you're looking for?" We all burst out laughing with delight. During their mapping, they had recognized the spherule bed as a peculiar marker precisely at the KT boundary and had traced it across much of the states of Nuevo León and Tamaulipas, wondering what it could possibly be. They knew where it was and we knew what it was—it was the perfect combination.

Over the next few days we studied one KT outcrop after another in superb semidesert exposures—El Mulato where the marker-bed cliff slants up across a muddy hillside, El Peñon where you can wander across the exposed top of the KT sandstone bed over an area as big as two or three football fields, Cuauhtémoc where the spherules fill a deep channel gouged into the Cretaceous mud by the tsunami, and Rancho Nuevo where the heavy KT sand had sunk into the soft mud under-

Searching for the KT boundary in northeastern Mexico in January 1992. *From left*: Ricardo Martinez, Pedro Romero, Eduardo Ruiz, Manuel Grajales, Jan Smit, Nicola Swinburne, Milly Alvarez, Walter Alvarez.

neath. Every evening over dinner, Ricardo, Pedro, and Eduardo would tell us about the geology and the history of northeastern Mexico, and we would tell them about the KT mass extinction and the search for the site of the impact. After studying nine new outcrops we ran out of time, said goodbye to our PEMEX colleagues in a freak snowstorm, and flew home laden down with new samples to analyze.

THE TURNING POINT

It seemed like the turning point, that winter of 1991–92. Jan Smit and I had been convinced since 1980 that an impact had killed the dinosaurs. For more than ten years the evidence had looked better and better for a KT impact,[23] but there were always serious questions and nagging doubts, and there had been so many hopes dashed and so many frustrations in trying to find the impact site. Now there was a convincing crater at Chicxulub, and melt-rock samples from the crater, and glass from the KT boundary at Mimbral and Beloc and Leg 77. Laboratory results were coming in, leaving little room for doubt that the KT boundary glass came from Chicxulub, or that Chicxulub was one of the largest impact craters on Earth.

Our second trip to Mexico had the same flavor—the sense that all the pieces were falling into place, and that at last we really did understand what had happened in and around the Gulf of Mexico at the end of the Cretaceous.[24] On our previous trip, the outcrop at Arroyo el Mimbral had come like a gift, after days of fruitless searching and growing pessimism. On our second Mexican trip, everything was different. We knew what to look for, and day after day we found new outcrops just where we expected to find them, each one telling us more about what had happened on that terrible day 65 million years ago.

It was the tsunami that gave the mystery away. After years of frustration the tsunami deposits had finally led the detectives to the scene of the crime, and at last everything was coming together. For me personally, the symbolic turning point came when Ricardo Martinez, Pedro Romero, and Eduardo Ruiz showed those spherules, all full of bubbles, to Jan and me in the plaza at General Terán and led us out to see the tsunami bed at the KT boundary, all across Tamaulipas and Nuevo León.

CHAPTER 7

The World after Chicxulub

THE DAWN OF THE CENOZOIC

The Cretaceous-Tertiary boundary marks a profound discontinuity in Earth history. The early geologists were right to choose it as the dividing line to separate fundamental eras in the history of life—the Mesozoic and the Cenozoic—the era of Middle Life and the era of Recent Life. After the impact at Chicxulub, 65 million years ago, life on Earth was changed forever. The long-standing and stable reign of the dinosaurs had been destroyed by a chance event. The new world was inherited by a different cast of characters, and the previously insignificant mammals came to dominate life on the land.

It is worth pondering the realization that each of us is descended from unknown ancestors who were alive on that day when the fatal rock fell from the sky. They survived and the dinosaurs did not, and that is the reason why we are here now—as individuals and as a species. That one terrible day undid the benefits which 150 million years of natural selection had conferred upon the dinosaurs, making them ever fitter to be the large land animals of Earth. Evolution had not equipped them to survive the environmental disasters inflicted by a huge impact, and when the holocaust was over, they were gone.

Evolution had not provided impact resistance for the mammals either, but somehow they did survive. No one knows why, but it must have helped that they were smaller and therefore much more numerous than the dinosaurs, so that there was a better statistical chance that some would live.

When the environmental disruptions from the impact had waned and the mammal survivors emerged into a new world, they must have faced great dangers and great opportunities. Every species evolves to make its living in a specific way—to occupy a specific ecological niche. Some niches must have been eliminated by the extinction, in the sense that any mammal species dependent in any way upon dinosaurs, or which depended on a plant that disappeared, would also become extinct. Other niches must have opened up, and the most notable are the niches for large land animals. Before their extinction, the dinosaurs held possession of these niches, and all mammals were small. But one of the most remarkable features of mammal evolution after the KT extinction was the rapidity with which large land mammals evolved. In addition, the *number* of mammal species quickly went up, as mammals evidently found all kinds of new niches—new ways to exploit the world around them.

STUDYING THE CHICXULUB CRATER

From our perspective as human beings, one of the most important events in life history has been the emergence of human intelligence and its manifestations—language, writing, civilization, science, technology, and the arts. This has been the unprecedented achievement of a single species of mammals, *Homo sapiens.* It has happened at a rate which is breathtaking from the geological point of view. The 35,000 years of advanced human culture seems a long time compared to our individual life spans, but in geological terms it is the trivial interval of 0.035 million years! What we will do with our new capabilities is not yet clear, but as the twentieth century winds down, perhaps the most vibrant intellectual activity of our species is international science—the global endeavor to understand the Cosmos, the planet we live on, life in its nearly endless varieties, and the laws of Nature which underlie everything we can see and study.

Our particular scientific quest reached a turning point with the discovery of the crater at Chicxulub, because the 10-year search for the impact site was over. But since then, questions we could not previously tackle have come to the fore. By studying the Chicxulub Crater and its surroundings, we can now investigate what happens in a giant impact, in the very unfamiliar regime of velocities that cannot be duplicated in the laboratory, and of shock waves and temperatures completely beyond normal experience.

This puts us in a better position to approach the most difficult question—What kind of environmental disruption caused the disappearance of each group of plants and animals that became extinct? It is hard to imagine any direct physical evidence that could ever reveal what killing mechanism finished off *T. rex*, when we are unlikely ever to see a specimen of *T. rex* that was alive at the moment of the impact. Nevertheless, we can speculate about killing mechanisms in a more intelligent way, knowing that the impact site was underlain by limestone and anhydrite. That kind of target must have released vast amounts of carbon dioxide and sulfur. If the impact had been in the ocean, it would have produced enormous quantities of water vapor, whereas impact on granite or metamorphic rock would have released relatively little of any of these gases.

So now geologists and geophysicists are concentrating much effort on studying the Chicxulub Crater and its surroundings, and Mexican scientists are taking the lead. Perhaps the most exciting approach to studying the crater itself is the shallow drilling program of UNAM, the National University in Mexico City, led by Luis Marín and Jaime Urrutia. Deep drilling is so expensive that it will take a long time to plan and finance, but in the meanwhile, with a small drilling rig and Mexican funding, the UNAM scientists have been able to reach the top of the buried ejecta blanket. The UNAM wells have recovered the first new impact material, and the cores will be made available to the international research community.

Geophysicists have indirect ways of studying buried structures, and these are being used intensely. By studying slight variations in the pull of gravity, a group led by Buck Sharpton, at the Lunar and Planetary Institute in Houston, reports ring-shaped features out to a diameter of 300 km.[1] But Alan Hilde-brand's group, using the same gravity methods, finds no rings beyond 170 km diameter.[2] This difference has led to some heated disputes at scientific meetings.

More detailed information on buried structures comes from seismic reflection studies. For a long time the best seismic data available came from lines recorded years ago by Dick Buffler of the University of Texas. Two new PEMEX seismic lines crossing the crater just north of the coast have now been published by Antonio Camargo and Gerardo Suárez, the geophysics profes-sor who is now the Provost for Research of UNAM,[3] and more seismic profiling is under way.

All these methods of studying the crater go hand in hand. The seismic lines provide single traverses across the crater showing the pattern of features at depth. Drill holes make it possible to identify and interpret the features seen on the seismic lines. And finally, gravity measurements, collected over the whole area, allow geophysicists to extend the two-dimensional seismic in-formation to give a complete, three-dimensional picture of the crater.

These usual geophysical approaches were supplemented by a completely unexpected line of evidence discovered by Kevin Pope, Adriana Ocampo, and Charles Duller. Kevin is a consult-ing geologist and archaeologist in Los Angeles who has years of experience in the Yucatán, his wife Adriana is a planetary scien-tist at the Jet Propulsion Laboratory in Pasadena and an expert in remote sensing of planetary surfaces, and Charles Duller is at NASA Ames Research Center near San Jose. They plotted on a map of the Yucatán the distribution of cenotes—the small, round, spring-fed lakes which provided the fresh water that made the Mayan civilization possible. To their surprise they found that the cenotes fall on a nearly perfect circular ring that outlines the buried Chicxulub Crater. We had been wrong in thinking there was no trace of the crater at the surface. No one yet understands how a deeply buried crater can control the pat-tern of springs far above it, but the cenote ring is unmistakable.[4]

THE SEARCH FOR THE CLOSEST OUTCROP

Meanwhile, there has been an intensified search for KT outcrops as close as possible to the crater. The deposit of ejecta should get very thick, and the fragments larger and larger, within a couple of hundred kilometers of the crater rim. The Mexican State of Chiapas seemed like a good place to look, despite the tropical vegetation cover and civil unrest. Mexican geologists took the lead in Chiapas—especially Juan Bermudez, along with Manuel Grajales and his wife, paleontologist Maria del Carmen Rosales—and Philippe Claeys and Sandro Montanari went with them on some of their trips. Luis Marín and Buck Sharpton have also studied the boundary in Chiapas, and so has Haraldur Sigurdsson. The KT outcrops in Chiapas are unique. They seem to record the collapse of the edge of a shallow-water limestone platform, with fragments of limestone that were shaken loose, or washed away, ending up in the adjacent deep water, just as the impact debris was falling.

I think most of us doubted whether a real outcrop of the ejecta blanket would ever be found, because the Yucatán has been subsiding for more than 150 million years and it seemed likely that not only the crater but all of the surrounding ejecta blanket would be buried by younger sediments. But Kevin and Adriana's discovery of the cenote ring encouraged them to search for outcrops in the vast tropical plains of the Yucatán. They criss-crossed the peninsula, sometimes with Al Fischer, looking at every hill that broke the flat horizon and examining every quarry they could find. The fruit of this search was their discovery of one of the most exciting, puzzling KT outcrops ever found. In January of 1995, Philippe, Milly, and I went on an expedition sponsored by The Planetary Society, with Kevin and Adriana, Eugene Fritsche from California State University at Northridge, and Mexican paleontologist Francisco Vega, to study this locality.

The little country of Belize lies tucked in the southeast corner of the Yucatán Peninsula, between Mexico and Guatemala.

Adriana Ocampo and Kevin Pope at the Albion Island quarry in Belize, where they discovered the closest outcrop of KT boundary ejecta to the Chicxulub Crater.

Driving west from Orange Walk, we came to the village of San Antonio. Here the Rio Hondo divides and flows along both sides of Albion Island, where there is a low hill flanked by cenotes and surrounded by flat plains. The inside of the hill is freshly exposed in the walls of a quarry. The lower 25 meters of exposure in the quarry is evenly layered dolomite, the characteristic bedrock of this part of the Yucatán. Dolomite forms when half of the calcium in limestone is replaced by magnesium, a process which usually erases most of the historical information originally held by the limestone. Fossils are commonly destroyed when limestone is altered to dolomite, making it very difficult to date these rocks. But Francisco, with his keen paleontologist's eye, was able to find fossils of crabs—his particular specialty—and of snails. After discussions between Francisco and Jan Smit, the fossil snails were dated as very late Cretaceous, which increased our interest in what lay on top of the dolomite in the Albion Island quarry.

Above the Cretaceous bedrock is a 15-meter deposit of dolomite fragments. Kevin and Adriana believed that this was the ejecta blanket from the Chicxulub Crater, but it didn't look right to Philippe and me. A crater that large should have excavated deep into the Yucatán crust, throwing out blocks of all kinds, including granite from the continental crust, but the fragments at Albion Island are almost exclusively dolomite—the surface layer in the target. In addition, the blocks thrown out of a crater should be angular. Geologists use the Italian word *breccia* for a deposit of angular blocks, and that is what we expected the ejecta blanket would look like, but the dolomite fragments in the quarry are rounded. At first these observations led Philippe and me to doubt if this was really the ejecta blanket. Day after day we examined the rock exposures in the quarry and debated whether this was Chicxulub ejecta or not. Eventually we were largely convinced of an impact origin because of the presence, at the base of the fragment layer, of little clay objects, unique among all the dolomite debris, which appear to be altered droplets of glass.

If the Albion Island deposit really is the Chicxulub ejecta blanket, then its unexpected characteristics provide new information on the impact event. Perhaps the fragments come not from the main crater, but from secondary craters, where really large ejecta blocks fell, making their own small craters, which only penetrated the shallow dolomite. And perhaps the fragments reached Belize as a ground-hugging flow of gas and water and rocks, so that the dolomite fragments were rounded by abrasion during transport. Or perhaps the explanation is completely different. Laboratory studies of the samples we took should provide the answer. Already Frank Asaro has found anomalous iridium at the very base of the ejecta deposit, indicating that it contains material derived from the impacting object. Bruce Fouke, an expert in the study of limestone and dolomite who has come to Berkeley as a research scientist, is finding a remarkably detailed history of events before, during, and after the impact, recorded in the Albion Island dolomite. And as the Albion Island material is examined in the lab, more and more strange features are emerging, promising new understanding of the Chicxulub event.

THE DOUBLE FIREBALL FROM CHICXULUB

Once we knew the location of the crater, we could begin to think about how the ejecta had been dispersed around the world. So I started making calculations of the ballistic trajectories of the impact ejecta. The Chicxulub fireball would be big enough to blow the ejecta clear through the atmosphere and launch the particles on ballistic trajectories which would end at the points all over the world where the ejecta was deposited. The pattern of falling ejecta would be simple on a slowly rotating body like the Moon, but I found that on the more rapidly spinning Earth, the pattern is complicated and not at all what you might expect. The Earth rotates beneath the ejecta while it is still aloft, and as a result the ejecta lands to the west of where it was aimed.[5]

One day Philippe looked over my shoulder at a map showing the calculated deposit of ballistic ejecta, and said, "Look how the steep ejecta from Chicxulub falls in the Pacific Ocean! I'll bet that explains the shocked quartz Jennifer has been finding." Jennifer Bostwick is a graduate of our department who went to UCLA for graduate work under the guidance of Frank Kyte and John Wasson. Studying the KT boundary in sediment cores from the Pacific Ocean, she and Frank had made a remarkable discovery—shocked quartz grains are very abundant there, whereas the same distance away from Chicxulub to the east, shocked quartz is nearly absent. This asymmetry in the pattern of shocked quartz was completely unexpected.[6]

But the observed asymmetry looked just like the calculated map on our computer screen, and suddenly it all made sense—the Earth's rotation had distorted the pattern of the falling shocked quartz. In order for this to happen, the shocked quartz grains would have to be launched on steep trajectories. Philippe and I quickly realized that if the quartz grains took off at 70° up from horizontal and the melted ejecta drops took off at 45°, the droplets would arrive in places like Montana before the shocked quartz, and this would explain the puzzling double boundary layer mentioned in chapter 5. The double layer[7] is a characteristic of the KT boundary in the western United States, with the shocked quartz grains lying just above the spherules which were derived from droplets of impact melt. The separation is so sharp that it had seemed to suggest two impacts at the KT boundary, slightly separated in time. We could explain both layers as coming from Chicxulub if the quartz took off more steeply than the impact melt. But why would that happen?

We needed to talk with an expert in the dynamics of impact events, and fortunately Susan Kieffer visited Berkeley just then. Sue has studied all kinds of fast moving geologic processes—vigorous rapids like Lava Falls in the Grand Canyon; volcanic explosions, where she explained what happened when Mount St. Helens blew up; geyser eruptions, where she explored Old Faithful by lowering a robot down the vent; and impact cratering, starting with her Ph.D. thesis on Meteor Crater supervised by Gene Shoemaker.[8] Sue is a fine musician, and she once told me that the slow passages labeled *adagio* always bored her—she likes her music *presto* or *allegro molto vivace*, and she likes geologic processes that move fast.

Philippe and I told Sue that we could explain both Jennifer's shocked quartz pattern and the double ejecta layer if there was some way to launch the quartz on steeper trajectories than those of the melt droplets. After a day of thinking about it, Sue came up with an explanation that seems to work. In an impact on most kinds of rocks, the solid and liquid ejecta are launched on roughly 45° trajectories, forming an expanding, ringlike "ejecta curtain." This ejecta curtain is separate from the fireball or "vapor plume," a cloud of vaporized impactor and target rock emitted from the site of the impact. These normal impacts are understood in considerable detail.[9] But Sue realized that the Chicxulub target was very unusual, and all that limestone must have released a huge cloud of CO_2 during an impact. There must have been not one, but two gaseous fireballs—the first a cloud of extremely hot, vaporized rock, and the second a cloud of CO_2 vapor at a less elevated temperature, given off by more lightly shocked limestone. Sue's calculations gave just the dynamics Philippe and I had inferred from the pattern of the ejecta. It was satisfying to see details of the impact event fall into place so neatly.[10]

GEOLOGISTS IN THE POST-UNIFORMITARIAN WORLD

Chicxulub marked a watershed. With the KT crater found at last, the kind of hard-core uniformitarianism which automatically rejects all inferences of catastrophic events was dead. Though no serious scientist doubts that *most* Earth change is gradual, geologists are now free to explore the occasional catastrophic events which have punctuated Earth history.

There is a striking asymmetry between the plate tectonics revolution and the change in view required by the KT impact theory. The plate tectonics revolution was uniformitarian in concept, but it dramatically changed the lives of almost all geologists because the evidence for plate tectonics is everywhere. Plate tectonics processes have gone on continuously for at least a thousand million years, leaving their imprint on almost every aspect of Earth history recorded in rocks. Our science was completely transformed by plate tectonics.

Ironically, the acceptance of impacts and the death of hardcore uniformitarianism, although catastrophic in concept, has had a more gradual and gentle effect on geologists. Impacts have been so rare that it is difficult to find evidence about them. Except for the largest impacts, it is hard even to find their positions in the stratigraphic record. A great impact like the KT boundary event, large enough to cause a mass extinction, can easily be located in the rock record, because the fossils are different above and below the impact level. Searching for smaller impacts that did not have a dramatic effect on life is very difficult, and their deposits are more likely to be found by chance.

Nevertheless, slow progress is being made, and geologists now know of several deposits of impact ejecta in the stratigraphic record in different parts of the world. This kind of evidence for impact supplements the list of known impact craters,[11] which has now reached about 130. The ejecta levels in the stratigraphic record span the age range from extremely ancient to very recent. Some are associated with mass extinctions; others were due to impacts too small to have more than local biologic effects.

Don Lowe at Stanford, working on sedimentary rocks of the Precambrian, deposited long before the appearance of abundant fossils, was intrigued by a bed of spherules, and was able to show that they are ejecta from an ancient impact.[12] In Australia, Victor Gostin and his colleagues found a layer of impact ejecta in a Precambrian sedimentary sequence and were able to show that it came from the Acraman impact crater about 300 km away.[13]

There are almost no fossils, and thus no evidence of biological extinction, in Precambrian strata, but the detailed fossil record of the 570 million years since the end of the Precambrian gives evidence of five great mass extinctions and about five smaller ones. The KT boundary is the most recent of the five great extinctions and has yielded much more information than any of the others. Early in the KT work, our Berkeley group imagined that all mass extinctions were caused by impacts. That may still be the case,[14] but it is important to emphasize that nothing like the panoply of impact evidence at the KT boundary has been found for any other mass extinction.

Nevertheless, tantalizing bits of evidence do exist. A major extinction occurred at the boundary between the Frasnian and Fammenian stages, near the end of the Devonian Period, about 365 million years ago. It was so abrupt that Canadian paleontologist and geologist Digby McLaren, in his 1970 Presidential Address to the Paleontological Society of America,[15] suggested that this extinction might have been caused by an impact event. Ten years ahead of its time, McLaren's suggestion was totally ignored, but recently impact-glass spherules have been found at the Frasnian-Fammenian boundary in China by Kun Wang[16] and in Belgium by Jean-Georges Casier and Philippe Claeys.[17] Digby McLaren is now seen as a prophet.

Another of the great extinctions, at the Triassic-Jurassic boundary, 205 million years ago, gave evidence of an impact origin when Dave Bice and Cathy Newton found shocked quartz grains at that level in an outcrop in Italy.[18] One of the smaller, less abrupt extinction events, near the Eocene-Oligocene boundary 34 million years ago, which has been intensively studied by a group led by Sandro Montanari, has yielded shocked quartz and anomalous iridium, giving evidence for more than one impact.[19] Two large craters of that age have been found, one in Siberia and the other under Chesapeake Bay.[20]

Other stratigraphic evidence for impact has turned up, by accident, at levels not marked by extinctions. Frank Kyte has worked extensively on ejecta from a young impact (Pliocene, dating from 2.3 million years ago) that he discovered in ocean bottom sediments in a remote part of the South Pacific.[21]

The most spectacular impact deposit of all was discovered in the mountains around Alamo, in southern Nevada, by John Warme and his students from Colorado School of Mines. While studying what seemed to be normal, bedded limestones of Devonian age, they gradually came to realize that enormous blocks of bedrock the size of office buildings had broken loose, and slurries of limestone fragments had been injected in underneath them. On top of this array of huge, slightly displaced blocks, they discovered a breccia deposit, made of angular fragments ripped up from the shallow sea bottom which covered this part of Nevada in the Devonian. From the blocks to the breccia, the Alamo deposit gives the impression of a sudden, violent disruption of the sea floor, tearing loose the shallowest part of the bedrock and prying up huge portions of the deeper strata. John named this remarkable collection of blocks and fragments the Alamo Breccia, and has invited many geologists to visit the area with him as he carefully considered whether the obvious violence it records was the result of impact or some other catastrophic event. Finally he found unmistakable grains of shocked quartz within the breccia, and a link to an impact was established.[22] John's current thinking is that a large impact in the ocean to the west produced a huge tsunami which caused this damage to the sea floor when it crashed into the continental margin in Nevada. For a while it seemed that this impact might explain the Frasnian-Fammenian mass extinction of the Late Devonian, but careful paleontological dating by Charles Sandberg has shown that the Alamo Breccia impact is about 3 million years older than the mass extinction.[23] It seems to have been an impact big enough to have inflicted spectacular damage on the continental margin, but not big enough to have disrupted the global biosphere.

The record of impact events preserved in stratified sediments is meager, and at present far more craters are known than ejecta deposits. But we can expect the stratigraphic record of impacts to grow, as more and more geologists learn about impacts and are thus in a position to recognize ejecta deposits when they come across them.

A ROLE FOR VOLCANISM?

Impact as a geologic process was long ignored by geologists. Now it must be recognized as a rare but significant kind of event, and evidently the cause of at least the KT mass extinction. Volcanism has always been of prime interest to geologists, and it was brought forward by some geologists as an explanation for the KT extinction. Now that there is such a strong case for impact as the cause of that biological turning point, can volcanism be dismissed from the list of catastrophic events with global effects?

Not yet, it seems, for there remain some intriguing but mystifying hints that volcanism is somehow involved. As mentioned in chapter 5, the huge volcanic province of the Deccan Traps[24] in India, proposed by Dewey McLean as the cause of the KT extinctions, has been dated by Vincent Courtillot's group as falling extremely close to the KT boundary. Yet the many lava flows with soil horizons between them indicate that Deccan volcanism went on too long to explain as brief an event as the KT extinction. I would have dismissed the apparent age match between the Deccan Traps and the KT impact-extinction event as a strange coincidence, if it were not that a second such coincidence has turned up.

The greatest of all the mass extinctions was at the Permian-Triassic boundary, 250 million years ago.[25] There is no evidence either for or against impact at that time, because there is very little preserved stratigraphic record across the Permian-Triassic boundary anywhere in the world. On the other hand, the greatest of all outpourings of lava on the continents is the Siberian Traps, much like the Deccan Traps but substantially larger in volume. Recently Paul Renne at the Berkeley Geochronology Center has obtained reliable dates on both the Siberian Traps and the Permian-Triassic boundary,[26] and they are indistinguishable!

A good detective shouldn't ignore even a single coincidence like the KT-Deccan match in timing, and when it is bolstered by a second coincidence like the match between the Siberian Traps and the Permian-Triassic boundary, it just has to be significant. But at the moment, I don't know of anyone with a reasonable explanation for a link between impacts, volcanism, and mass extinctions. The obvious idea—that impacts produce both volcanic eruptions and mass extinctions—seems unlikely because Chicxulub is nowhere near India.[27] Right now we are in a situation that scientists particularly enjoy—where there is an intriguing mystery, some obviously significant clues, and nobody has any idea what the explanation will be.

REENACTMENT

It is not yet clear what role volcanism will play in our final understanding of catastrophic events on Earth. But as the list of known impact craters increases by two or three a year, and as impact debris is found in more and more places in the rock record, the impact of comets and asteroids is being accepted by more and more geologists as a normal process on Earth, as it obviously is elsewhere in the solar system. The Yucatán impact was unusual because of its magnitude, which was sufficient to cause a mass extinction, but it is simply one of the larger events in a continuum of impact magnitudes. Objects of all sizes fall to Earth, and the smaller ones fall much more frequently. The smallest objects, of sand grain size and smaller, do not hit the Earth's surface because they burn up by friction high in the atmosphere, making the streaks of light we call meteors. Meteors are so frequent that almost anyone who lives away from bright lights can see one or two an hour in the dark night sky.

One would think that the only collisions visible in the sky would be the streaks of light from micrometeorites. But, unexpectedly, the dramatic event that fully confirmed the end of uniformitarian geology was not seen by looking down at the rocks which record Earth history, but by looking upward.

More than any other single person, Gene Shoemaker has been the central figure in the growing understanding of impact craters, in the expansion of geology throughout the solar system, and in the laying to rest of nineteenth-century uniformitarian dogma. From his early days of studying the Moon through a small telescope and dreaming of going there, to his proof that Meteor Crater is an impact scar, to his training of the astronauts for the lunar missions, to his scientific leadership of one deep-space probe after another, to his discoveries about the KT boundary impact, to his many expeditions to study impact craters in the Australian desert—Gene has always been at the forefront.

It is thus appropriate indeed that the final episode in this telling of the impact-extinction story should center on Gene Shoemaker, his wife Carolyn, and their friend David Levy.[28] For years Gene and Carolyn, often with David, would journey to the astronomical observatory on Palomar Mountain every month in the dark of the Moon, where they were systematically photographing the sky again and again, gradually building up a census of Earth-crossing asteroids—the space rocks whose orbits can come inside the orbit of the Earth, and which thus have a chance of hitting our planet. Gene wanted to know how many of these potential threats there are, how often on the average they hit the Earth, and whether there is any immediate danger for which we should be preparing.

Carolyn has the best eye for spotting asteroids and comets on the photographic plates, so it was she who called out to Gene and David, "Look at this—I think I've got a squashed comet!" Detailed pictures soon showed that the comet they had discovered was not squashed; it was fragmented. Periodic comet Shoemaker-Levy 9, as it was designated, had been captured by Jupiter so that it orbited the giant planet, rather than the Sun. On one pass, shortly before Carolyn noticed it, the comet came too close to Jupiter, gravitational forces ripped it into fragments, and dust drifting away from the surfaces of the freshly broken fragments made the pieces of the formerly inactive comet shine in the sunlight.

The orbits of the fragments were calculated, and to the surprise and delight of astronomers and geologists, it was clear that they were going to crash into Jupiter on their next pass. Observing programs were planned in feverish haste, and as the comet fragments bore down on Jupiter in July of 1994, telescopes of all kinds, all over the Earth and out in space, were trained on the site of the impending collisions. The impacts were even more spectacular than anyone had dared hope. As the larger fragments plunged into the nearly bottomless atmosphere of Jupiter, plumes of shocked material rose thousands of kilometers above the planet and then, in Jupiter's fierce gravity, collapsed on top of the atmosphere. Their collapse generated intense bursts of heat, which could be seen as infrared light through telescopes on Earth.

Nature was doing, at a safe distance, an experiment we could not possibly have done ourselves. Astronomers at their telescopes were awed by what they were seeing, and for those of us who had joined with Gene Shoemaker in the long fight to have impacts accepted by Earth historians, it was profoundly satisfying. It was the incontrovertible proof that big impacts are not simply a thing of the remote past. They can and do happen right now.

The bursts of heat from the impact sites on Jupiter were intellectually satisfying, but they were also sobering and deeply moving. For as we watched the violence being inflicted on another planet, we were seeing a reenactment of the last spectacle ever witnessed by *Tyrannosaurus rex*—the deadly flash from the Crater of Doom, on the day the Mesozoic world ended.

CHAPTER ONE
ARMAGEDDON

1. Meteorites are any pieces of fallen space debris that can be picked up on the ground. It is usually difficult to tell whether a particular meteorite came from an asteroid or from a comet.

2. For an explanation of how the diameter of the impactor was calculated, see Harte, J., 1988, Consider a spherical cow: Mill Valley, CA, University Science Books, 283 p. (p. 7–9).

3. The impact velocity must have been at least 11 km/sec—the Earth's escape velocity, which a rocket must reach in order to escape from Earth's gravitational pull. It is thus also the velocity gained by any object falling to Earth. But asteroids and comets start out with an initial velocity and thus would hit the Earth at more than 11 km/sec. An asteroid, starting fairly close to us, between Mars and Jupiter, and orbiting the Sun in the same direction as the Earth, would impact at about 20 km/sec. A comet, falling from the extreme outer fringes of the solar system and possibly orbiting the Sun in the opposite direction to the Earth, could have hit the Earth head on, with an impact velocity as high as 70 or 80 km/sec. The difference of a factor of 4, between 20 and 80 km/sec, would translate into a difference of a factor of 16 in the kinetic energy release, because kinetic energy is related to the square of the velocity (K.E. $= mv^2/2$). Since we do not know whether the impactor was an asteroid or a comet, the impact velocity remains uncertain and the energy of impact remains very uncertain. The impact velocity of 30 km/sec used here is chosen arbitrarily, as a value in the middle of the range of possible velocities.

4. A clear and thoughtful introduction to Nature's energy laws is given by Atkins, P. W., 1984, The second law: New York, Scientific American Library (W. H. Freeman), 230 p.

5. Melosh, H. J., Schneider, N. M., Zahnle, K. J., and Latham, D., 1990, Ignition of global wildfires at the Cretaceous/Tertiary boundary: Nature, v. 343, p. 251–254.

6. Wolbach, W. S., Gilmour, I., Anders, E., Orth, C. J., and Brooks,

R. R., 1988, Global fire at the Cretaceous-Tertiary boundary: Nature, v. 334, p. 665–669.

7. Bourgeois, J., Hansen, T. A., Wiberg, P. L., and Kauffman, E. G., 1988, A tsunami deposit at the Cretaceous-Tertiary boundary in Texas: Science, v. 241, p. 567–570.

8. Toon, O. B., Pollack, J. B., Ackerman, T. P., Turco, R. P., McKay, C. P., and Liu, M. S., 1982, Evolution of an impact-generated dust cloud and its effects on the atmosphere: Geological Society of America Special Paper, v. 190, p. 187–200.

9. Lewis, J. S., Watkins, G. H., Hartman, H., and Prinn, R. G., 1982, Chemical consequences of major impact events on Earth: Geological Society of America Special Paper, v. 190, p. 215–221; Pope, K. O., Baines, K. H., Ocampo, A. C., and Ivanov, B. A., 1994, Impact winter and the Cretaceous/Tertiary extinctions: results of a Chicxulub asteroid impact model: Earth and Planetary Science Letters, v. 128, p. 719–725.

10. Many paleontologists now refer to the extinct dinosaurs of the Mesozoic as the "non-avian dinosaurs," based on the evidence that birds are part of the dinosaur lineage.

11. Feduccia, A., 1995, Explosive evolution in Tertiary birds and mammals: Science, v. 267, p. 637–638.

12. Johnson, K. R. and Hickey, L. J., 1990, Megafloral change across the Cretaceous/Tertiary boundary in the northern Great Plains and Rocky Mountains, U.S.A.: Geological Society of America Special Paper, v. 247, p. 433–444.

13. Ward, P. D., A review of Maastrichtian ammonite ranges: Geological Society of America Special Paper, v. 247, p. 519–530.

14. Smit, J., 1982, Extinction and evolution of planktonic foraminifera after a major impact at the Cretaceous/Tertiary boundary: Geological Society of America Special Paper, v. 190, p. 329–352.

CHAPTER TWO
EX LIBRO LAPIDUM HISTORIA MUNDI

1. More precisely, the age of the Earth is 4,600 million years, but the round number 5,000 million years is close enough, and easier to remember for the purpose of appreciating Earth history. Usually the age of the Earth is stated as 4.6 *billion* years, but I find it helpful to think of

all events in Earth history in terms of a single fundamental unit of time—million-years.

2. It might seem logical to restrict the word "date" to numerical ages based on radioactive decay, and use some other word for the approximate time ranges determined from fossils. But geologists think of both fossils and radioactivity as ways to date rocks, and where necessary, distinguish the two as "fossil ages" and "numerical ages."

3. A good example comes from the Tertiary marine sediments above the KT boundary at Gubbio: Montanari, A., Drake, R., Bice, D. M., Alvarez, W., Curtis, G. H., Turrin, B. D., and DePaolo, D. J., 1985, Radiometric time scale for the upper Eocene and Oligocene based on K/Ar and Rb/Sr dating of volcanic biotites from the pelagic sequence of Gubbio, Italy: Geology, v. 13, p. 596–599.

4. The most recent comprehensive summary of the geologic time scale is by Harland, W. B., Armstrong, R. L., Cox, A. V., Craig, L. E., Smith, A. G., and Smith, D. G., 1990, A geologic time scale 1989: Cambridge, Cambridge University Press, 263 p.

5. One major challenge to *gradual* evolution has come from paleontologists Niles Eldredge and Stephen J. Gould, who find evidence for punctuated equilibrium—that individual species are stable for long intervals and that the appearance of new species takes place rapidly: Eldredge, N. and Gould, S. J., 1972, Punctuated equilibria: An alternative to phyletic gradualism, in Schopf, T.J.M., ed., Models in paleobiology: San Francisco, Freeman, Cooper and Co., p. 82–115; Gould, S. J., 1984, Toward the vindication of punctuational change, in Berggren, W. A., and Couvering, J.A.V., eds., Catastrophes and Earth history; The new uniformitarianism: Princeton, Princeton University Press, p. 9–34.

6. Luterbacher, H. P. and Premoli Silva, I., 1962, Note préliminaire sur une revision du profil de Gubbio, Italie: Rivista Italiana di Paleontologia e Stratigrafia, v. 68, p. 253–288.

7. The full history of research on magnetic reversals and plate tectonics is told by Glen, W., 1982, The road to Jaramillo: Stanford, Stanford University Press, 459 p.

8. Upper Cretaceous-Paleocene magnetic stratigraphy at Gubbio, Italy: Geological Society of America Bulletin, v. 88, 1977, p. 367–389. Part I: Arthur, M. A. and Fischer, A. G., Lithostratigraphy and sedimentology; Part II: Premoli Silva, I., Biostratigraphy; Part III: Lowrie, W. and Alvarez, W., Upper Cretaceous magnetic stratigraphy; Part IV:

Roggenthen, W. M. and Napoleone, G., Upper Maastrichtian-Paleo-
cene magnetic stratigraphy; Part V: Alvarez, W., Arthur, M. A.,
Fischer, A. G., Lowrie, W., Napoleone, G., Premoli Silva, I., and Roggenthen, W. M., Type section for the Late Cretaceous-Paleocene geomagnetic reversal time scale.

9. Lowrie, W. and Alvarez, W., 1981, One hundred million years of
geomagnetic polarity history: Geology, v. 9, p. 392–397.

10. The KT boundary clay layer was first pointed out by Luterbacher, H. P. and Premoli Silva, I., 1964, Biostratigrafia del limite creta-
ceo-terziario nell'Appennino centrale: Rivista Italiana di Paleontologia
e Stratigrafia, v. 70, p. 67–128, Fig. 3.

Chapter Three
Gradualist versus Catastrophist

1. Stephen J. Gould (1987, Time's arrow, time's cycle: Cambridge,
Harvard University Press, 222 p.) has carefully analyzed the writings
of Hutton and Lyell, seeing beyond the traditional picture of these
men, which he calls "portraits in textbook cardboard," and discovering a more complicated intellectual history, where the deeply held
philosophical views of Hutton and Lyell guided or even controlled the
kind of field observations they made.

2. Muir, J., 1894, The mountains of California: Century, New York,
chapter 1., reprinted in Muir, J., 1992, The eight wilderness discovery
books: Diadem, London, 1030 p., see p. 301. I have long thought of
Muir as a gradualist because of this passage, but recently Ken Deffeyes
pointed out to me that on another occasion, Muir was mistakenly a
catastrophist, interpreting talus slopes in the mountains as having
been shaken down by great earthquakes.

3. The discovery of the Alpine thrust faults is beautifully told by
E. B. Bailey, 1935/1968, Tectonic essays, mainly Alpine: Oxford, Oxford University Press, 200 p., especially chapter 4.

4. This approach to studying geologic processes is confusingly
called "actualism," from the Romance-language use of "actual" to
mean "present-time."

5. Stephen J. Gould (1965, Is uniformitarianism necessary?: American Journal of Science, v. 263, p. 223–228) has shown how Lyell, in his
very influential book (Lyell, C., 1830–33/1990–91, Principles of Geology,
reprint of the 1st edition: Chicago, University of Chicago Press, v. 1–3),

convinced geologists that all of Earth history was slow, gradual, and quiet. Throughout the *Principles*, Lyell used the word "uniformity" interchangeably, for two completely different concepts. ("Uniformity" was his term; "uniformitarianism" was coined later.) Lyell's first meaning for uniformity was the concept that the laws of Nature do not change from place to place, or as time passes. This is the common assumption of all scientists and the basis of the scientific method, and geologists have no quarrel with it. However, Lyell also used "uniformity" for a view of Earth history that was highly controversial among his contemporaries and which we now know was wrong. Lyell believed in uniformity of rate, or gradualism—the view that nothing in the Earth's past happened much faster than it is happening now, and in uniformity of state—the idea that conditions on the Earth in the past have always been about the same as they are now. By entangling uniformity of rate and state with the uncontroversial idea of uniformity of natural law, Lyell got these largely incorrect views accepted by his contemporaries and passed on to future geologists as the hallowed doctrine of uniformitarianism.

6. Lyell, C., 1830–33/1990–91, op. cit., v. 3, p. 328.

7. Browne, J., 1995, Charles Darwin: voyaging: Princeton, Princeton University Press, 605 p., esp. p. 186–190.

8. Quoted by Gould, S. J., 1982, The panda's thumb: New York, W. W. Norton, 343 p. (p. 179).

9. The story of Bretz and the great flood of eastern Washington is told by Gould, S. J., 1982, op. cit., chap. 19.

10. The role of Pardee is described by Baker, V. R., 1995, Surprise endings to catastrophism and controversy on the Columbia—Joseph Thomas Pardee and the Spokane Flood controversy: GSA Today, v. 5, p. 169–173.

11. Baker, V. R., The Spokane Flood controversy, chap. 1, p. 14, in Baker, V. R. and Nummedal, D., 1978, The Channeled Scabland: NASA Planetary Geology Program, Washington, DC, 186 p.

12. Baldwin, R. B., 1949, The face of the moon: Chicago, University of Chicago Press, 239 p.

13. Ursula Marvin first recognized and explored the uniformitarian character of plate tectonics and the way it overwhelmed the evidence for catastrophic impacts that was visible on planets and moons: Marvin, U. B., 1990, Impact and its revolutionary implications for geology: Geological Society of America Special Paper, v. 247, p. 147–154.

14. Wegener, A., 1929/1966, The origins of continents and oceans (translated from the German by J. Biram): New York, Dover, 246 p.

15. Cox, A., ed., 1973, Plate tectonics and geomagnetic reversals: San Francisco, W. H. Freeman, 702 p.; Allègre, C. J., 1988, The behavior of the Earth: continental and sea-floor mobility: Cambridge, MA, Harvard University Press, 272 p.

16. Cox, A. and Hart, R. B., 1986, Plate tectonics: how it works: Palo Alto, CA, Blackwell, 392 p.

17. Winchell, A., 1886, Walks and talks in the geological field: New York, Chautauqua Press, 329 p. (p. 252).

18. Schuchert, C. and Dunbar, C. O., 1933, A textbook of geology, part II—Historical Geology (3rd edition): New York, Wiley, 551 p. (p. 381).

19. In a careful analysis a few years later, paleontologists Phil Signor and Jere Lipps demonstrated that for a truly abrupt extinction, the poorer the fossil record the more gradual the extinction appears. This came to be called the "Signor-Lipps effect": Signor, P. W. and Lipps, J. H., 1982, Sampling bias, gradual extinction patterns, and catastrophes in the fossil record: Geological Society of America Special Paper, v. 190, p. 291–296.

20. Gartner, S. and Keany, J., 1978, The terminal Cretaceous event: a geologic problem with an oceanographic solution: Geology, v. 6, p. 708–712.

21. Terry, K. D. and Tucker, W. H., 1968, Biologic effects of supernovae: Science, v. 159, p. 421–423.

22. Russell, D. and Tucker, W., 1971, Supernovae and the extinction of the dinosaurs: Nature, v. 229, p. 553–554.

CHAPTER FOUR
IRIDIUM

1. Alvarez, L. W., Anderson, J. A., El Bedwei, F., Burkhard, J., Fakhry, A., Girgis, A., Goneid, A., Hassan, F., Iverson, D., Lynch, G., Miligy, Z., Moussa, A. H., Sharkawi, M., and Yazolino, L., 1970, Search for hidden chambers in the pyramids: Science, v. 167, p. 832–839.

2. Rich has told the story in considerable detail: Muller, R. A., 1988, Nemesis: the death star (The story of a scientific revolution): New York, Weidenfeld and Nicolson, 193 p.

3. The six platinum-group elements (ruthenium, rhodium, palladium, osmium, iridium, and platinum) are clustered on the periodic table. All are rare in the solar system and they show rather similar chemical behavior.

4. One can never measure an absolutely accurate concentration, so the results are reported together with their analytical uncertainties. For example, an iridium value of 20±5 ppb (2σ), which is read "20 plus or minus 5 ppb, at 2 standard deviations uncertainty," means Frank is 95% confident that the true value lies between 15 and 25 ppb.

5. Years later we discovered that one whole batch of samples had been contaminated by the tiny amount of iridium in the platinum wedding ring of the technician who prepared them!

6. Sarna-Wojcicki, A. M., Morrison, S. D., Meyer, C. E., and Hillhouse, J. W., 1987, Correlation of upper Cenozoic tephra layers between sediments of the western United States and eastern Pacific Ocean and comparison with biostratigraphic and magnetostratigraphic age data: Geological Society of America Bulletin, v. 98, p. 207–223.

7. Frank Asaro has explained in detail the considerations that went into making the first iridium measurements and how they were done. See Asaro, F., 1987, The Cretaceous-Tertiary iridium anomaly and the asteroid impact theory, p. 240–242, in Trouwer, W. P., ed., Discovering Alvarez—Selected works of Luis W. Alvarez with commentary by his students and colleagues: Chicago, University of Chicago Press, 272 p.

8. Terry, K. D. and Tucker, W. H., 1968, Biologic effects of supernovae: Science, v. 159, p. 421–423; Russell, D. A. and Tucker, W., 1971, Supernovae and the extinction of the dinosaurs: Nature, v. 229, p. 553–554; Ruderman, M. A., 1974, Possible consequences of nearby supernova explosions for atmospheric ozone and terrestrial life: Science, v. 184, p. 1079–1081.

9. The most recent "nearby" supernova occurred in a galaxy adjacent to the Milky Way in 1987, and led to a greatly increased understanding of stellar explosions. The scientific story of the 1987 supernova is told by Dauber, P. M. and Muller, R. A., 1996, The three big bangs: New York, Addison-Wesley, 207 p.

10. For plutonium-244, about 55 half-lives of 83 Myr have passed since the formation of the Earth about 4,600 million years ago. During each half-life, half of the atoms present at the beginning of that 83 Myr period decay, so at the end of 55 half-lives the plutonium-244 will be

reduced to $1/(2^{55})$, or about 3×10^{-17} of its original abundance, which is far below the level we would expect from a KT supernova, and essentially undetectable.

11. Helen's last name is pronounced "Michael."

12. After a great deal of effort, Frank and Helen traced the impurity in the first sample to an experiment in another laboratory nearby. The impurity was at such a low level that it could only be detected by an extremely sensitive technique like NAA: see Asaro, F., 1987, The Cretaceous-Tertiary iridium anomaly and the asteroid impact theory, in Trouwer, W. P., ed., Discovering Alvarez: selected works of Luis W. Alvarez, with commentary by his students and colleagues: Chicago, University of Chicago Press, p. 240–242; Alvarez, L. W., 1987, Alvarez—Adventures of a physicist: New York, Basic Books, 292 p.

13. This compilation showed 2,561 genera during the 5–10 Myr before the KT event and 1,392 genera during the 5–10 Myr after it, for a survivorship of 54% or less, depending on how many Tertiary genera were newly evolved: see p. 41–42 in Russell, D. A. and Rice, G., eds., 1982, K-TEC II—Cretaceous-Tertiary extinctions and possible terrestrial and extraterrestrial causes: Syllogeus, National Museums of Canada, Ottawa no. 39, 151 p.

14. Sepkoski, J. J., Jr., 1982, A compendium of fossil marine families: Milwaukee, Milwaukee Public Museum, 125 p.; Raup, D. M. and Sepkoski, J. J., Jr., 1982, Mass extinctions in the marine fossil record: Science, v. 215, p. 1501–1503; Raup, D. M. and Sepkoski, J. J., Jr., 1983, Mass extinctions in the fossil record: Science, v. 219, p. 1239–1241.

15. Bob Dietz was a known heretic. It was only years later that he received the Penrose Medal—the highest honor of the Geological Society of America—in recognition of his having been right about plate tectonics and impact cratering at a time when most geologists considered both theories anathema.

16. The intense controversy over the origin of Meteor Crater is recounted by Hoyt, W. G., 1987, Coon Mountain controversies—Meteor Crater and the development of impact theory: Tucson, University of Arizona Press, 442 p. The Meteor Crater debate offered a preview of the coming battle over the KT impact hypothesis.

17. Walter Bucher was the strongest proponent of an internal cause for these craters: Bucher, W. H., 1963, Cryptoexplosion structures caused from without or from within the Earth? ("astroblemes" or "geoblemes"?): American Journal of Science, v. 261, p. 597–649. This paper by Bucher was followed by a discussion paper making the case

for an impact origin—Dietz, R. S., 1963, Cryptoexplosion structures: a discussion: American Journal of Science, v. 261, p. 650–664.

18. Symons, G. J., ed., 1888, The eruption of Krakatoa, and subsequent phenomena: Royal Society, London, 494 p.

19. Alvarez, L. W., Alvarez, W., Asaro, F., and Michel, H. V., 1980, Extraterrestrial cause for the Cretaceous-Tertiary extinction: Science, v. 208, p. 1095–1108.

20. Smit, J. and Hertogen, J., 1980, An extraterrestrial event at the Cretaceous-Tertiary boundary: Nature, v. 285, p. 198–200.

21. Kyte, F. T., Zhou, Z., and Wasson, J. T., 1980, Siderophile-enriched sediments from the Cretaceous-Tertiary boundary: Nature, v. 288, p. 651–656.

22. Ganapathy, R., 1980, A major meteorite impact on the Earth 65 million years ago: evidence from the Cretaceous-Tertiary boundary clay: Science, v. 209, p. 921–923.

23. Orth, C. J., Gilmore, J. S., Knight, J. D., Pillmore, C. L., Tschudy, R. H., and Fassett, J. E., 1981, An iridium abundance anomaly at the palynological Cretaceous-Tertiary boundary in northern New Mexico: Science, v. 214, p. 1341–1343.

Chapter Five
The Search for the Impact Site

1. Glen, W., ed., 1994, The mass-extinction debates: How science works in a crisis: Stanford, Stanford University Press, p. 58.

2. Leon T. Silver, Kevin Burke, George Carrier, Lee Hunt, Heinz Lowenstam, J. Murray Mitchell, Robert Pepin, Peter H. Schultz, Eugene Shoemaker, and George Wetherill.

3. There have now been three Snowbird Conferences, dealing with impacts and mass extinctions, in 1981, 1988, and 1994, although the third one was actually held in Houston. Each conference has generated a major proceedings volume, and these are central sources of information on the development of research in this interdisciplinary area. The three volumes are: (1) Silver, L. T. and Schultz, P. H., eds., 1982, Geological implications of impacts of large asteroids and comets on the Earth: Geological Society of America, Special Paper, v. 190, 528 p.; (2) Sharpton, V. L. and Ward, P. D., eds., 1990, Global catastrophes in Earth history: An interdisciplinary conference on impacts, volcanism, and mass mortality: Geological Society of America, Special Paper,

v. 247, 631 p.; (3) Ryder, G., Fastovsky, D., and Gartner, S., eds., 1996, The Cretaceous-Tertiary event and other catastrophes in Earth history: Geological Society of America Special Paper, v. 307, 580 p.

4. Alvarez, W., 1991, The gentle art of scientific trespassing: GSA Today, v. 1, p. 29–34.

5. Clemens, E. S., 1994, The impact hypothesis and popular science: conditions and consequences of interdisciplinary debate, in Glen, W., op. cit., p. 92–120.

6. There have been far more papers of high quality than could possibly be mentioned in a short book. To those whose work I have omitted, my sincere apologies!

7. Wilford, J. N., 1985, The riddle of the dinosaur: New York, Knopf, 304 p.; Raup, D. M., 1986, The Nemesis Affair: New York, W. W. Norton, 220 p.; Hsü, K. J., 1986, The great dying: San Diego, Harcourt Brace Jovanovich, 292 p.; Lampton, C., 1986, Mass extinctions—One theory of why the dinosaurs vanished: New York, Franklin Watts, 96 p.; Alvarez, L. W., 1987, Alvarez—Adventures of a physicist: New York, Basic Books, 292 p.; Muller, R. A., 1988, Nemesis: the death star—The story of a scientific revolution: New York, Weidenfeld and Nicolson, 193 p.; Smit, J., 1990, Meteorite impact, extinctions and the Cretaceous-Tertiary boundary: Geologie en Mijnbouw, v. 69, p. 187–204; Alvarez, W., and Asaro, F., 1990, What caused the mass extinction? An extraterrestrial impact: Scientific American, v. 263 (October), p. 78–84; Courtillot, V. E., 1990, What caused the mass extinction? A volcanic eruption: Scientific American, v. 263 (October), p. 85–92; Raup, D. M., 1991, Extinction—Bad genes or bad luck?: New York, W. W. Norton, 210 p.; Officer, C. and Page, J., 1993, Tales of the Earth: paroxysms and perturbations of the blue planet: New York, Oxford University Press, 226 p.; Clement, F. and Clementi, D., 1993, Chi uccise i dinosauri?: Edimond, Città di Castello (PG), Italy, 199 p.; Glen, W., ed., 1994, The mass-extinction debates: How science works in a crisis: Stanford CA, Stanford University Press, 370 p.; Courtillot, V., 1995, La vie en catastrophes: Paris, Fayard, 278 p.; Vaas, R., 1995, Der Tod kam aus dem All: Stuttgart, Franckh-Kosmos, 208 p.; Dauber, P. M. and Muller, R. A., 1996, The three big bangs: New York, Addison-Wesley, 207 p.; Frankel, C., 1996, La mort des dinosaures: l'hypothèse cosmique: Paris, Masson, 172 p.; Officer, C. B. and Page, J., 1996, The great dinosaur extinction controversy: Reading, MA, Addison-Wesley, 209 p.; Archibald, J. D., 1996, Dinosaur extinction and the end of an era: New York, Columbia University Press, 237 p.

8. Clemens, W. A., Archibald, J. D., and Hickey, L. J., 1981, Out with a whimper, not a bang: Paleobiology, v. 7, p. 293–298; Clemens, W. A., 1982, Patterns of extinction and survival of the terrestrial biota during the Cretaceous/Tertiary transition: Geological Society of America Special Paper, v. 190, p. 407–413; Archibald, J. D. and Clemens, W. A., 1982, Late Cretaceous extinctions: American Scientist, v. 70, p. 377–385.

9. The stratigraphy in eastern Montana was unexpectedly difficult to interpret because of channels cut by the ancient rivers, as Jan Smit discovered when he went to study the area: Smit, J. and van der Kaars, S., 1984, Terminal Cretaceous extinctions in the Hell Creek Area, Montana: compatible with catastrophic extinction: Science, v. 223, p. 1177–1179.

10. Signor, P. W. and Lipps, J. H., 1982, Sampling bias, gradual extinction patterns, and catastrophes in the fossil record: Geological Society of America Special Paper, v. 190, p. 291–296.

11. Ward, P., Wiedmann, J., and Mount, J. F., 1986, Maastrichtian molluscan biostratigraphy and extinction patterns in a Cretaceous/Tertiary boundary section exposed at Zumaya, Spain: Geology, v. 14, p. 899–903; Ward, P. D., 1990, A review of Maastrichtian ammonite ranges: Geological Society of America Special Paper, v. 247, p. 519–530.

12. Orth, C. J., Gilmore, J. S., Knight, J. D., Pillmore, C. L., Tschudy, R. H., and Fassett, J. E., 1981, An iridium abundance anomaly at the palynological Cretaceous-Tertiary boundary in northern New Mexico: Science, v. 214, p. 1341–1343.

13. Thierstein, H. R., 1982, Terminal Cretaceous plankton extinctions: a critical assessment: Geological Society of America Special Paper, v. 190, p. 385–399; Surlyk, F. and Johansen, M. B., 1984, End-Cretaceous brachiopod extinctions in the chalk of Denmark: Science, v. 223, p. 1174–1177; Smit, J. and Romein, A. J. T., 1985, A sequence of events across the Cretaceous-Tertiary boundary: Earth and Planetary Science Letters, v. 74, p. 155–170; Nichols, D. J. and Fleming, R. F., 1990, Plant microfossil record of the terminal Cretaceous event in the western United States and Canada: Geological Society of America Special Paper, v. 247, p. 445–454; Johnson, K. R. and Hickey, L. J., 1990, Megafloral change across the Cretaceous/Tertiary boundary in the northern Great Plains and Rocky Mountains, U.S.A.: Geological Society of America Special Paper, v. 247, p. 434–444; Sheehan, P. M. and Fastovsky, D. E., 1992, Major extinctions of land-dwelling vertebrates at the Cretaceous-Tertiary boundary, eastern Montana: Geology, v. 20, p. 556–560; D'Hondt, S., Herbert, T. D., King, J., and Gibson, C., 1996,

Planktonic foraminifera, asteroids, and marine production: Death and recovery at the Cretaceous-Tertiary boundary, in Ryder, G., Fastovsky, D., and Gartner, S., eds., The Cretaceous-Tertiary event and other catastrophes in Earth history: Geological Society of America Special Paper, v. 307, p. 303–317; Pospichal, J. J., 1996, Calcareous nannoplankton mass extinction at the Cretaceous/Tertiary boundary: an update, in Ryder, G., Fastovsky, D., and Gartner, S., eds., op. cit., p. 335–360. Huber, B. T., 1996, Evidence for planktonic foraminifer reworking versus survivorship across the Cretaceous-Tertiary boundary at high latitudes, in Ryder, G., Fastovsky, D., and Gartner, S., eds., op. cit., p. 319–334.

14. Foraminiferal paleontologist Gerta Keller at Princeton is now the strongest proponent of the viewpoint that the KT extinctions did not result from impact. For a collection of detailed paleontological papers largely favoring that view, see MacLeod, N. and Keller, G., eds., 1996, Cretaceous-Tertiary mass extinctions: biotic and environmental changes: New York, W. W. Norton, 575 p. See also Archibald, J. D., op. cit.

15. Grieve, R. A. F., 1982, The record of impacts on earth: implications for a major Cretaceous/Tertiary impact event: Geological Society of America Special Paper, v. 190, p. 25–37.

16. The distinction between rocks and minerals is important. The solid Earth is made up of rocks and, in turn, rocks are composed of minerals. Minerals are crystal grains with a characteristic chemical composition and atoms arranged in precisely ordered geometrical patterns. Rocks are made up of mineral grains, but the amount of each mineral can vary, so the chemical compositions of rocks are more variable than those of minerals. For example, the minerals quartz and calcite have the precise chemical compositions of SiO_2 and $CaCO_3$, respectively. But rocks made of mixtures of those two minerals can vary from a quartz sandstone with just a bit of calcite, through an evenly mixed sandy limestone, to a limestone that is almost all calcite, with just a few grains of quartz.

17. The chemical formulas for quartz, SiO_2, and for carbon dioxide, CO_2, look very similar, although their structure and properties are completely different. Quartz occurs as solid mineral grains with enormous numbers of atoms organized in a repeating crystalline structure, with a *ratio* of two oxygen atoms for each silicon atom. Carbon dioxide is a gas, in which each independent molecule has just three atoms—one carbon and two oxygens. In all mineral formulas, the sub-

scripts show the proportions among the vast numbers of atoms in the mineral grain.

18. Smit, J. and Klaver, G., 1981, Sanidine spherules at the Cretaceous-Tertiary boundary indicate a large impact event: Nature, v. 292, p. 47–49.

19. The different isotopes of a particular element have the same number of protons in their nucleus, but they have different numbers of neutrons and thus different weights. Carbon-12 has 6 protons and 6 neutrons, whereas carbon-13 has 6 protons and 7 neutrons, and both are stable. Carbon-14 has 6 protons and 8 neutrons, and decays radioactively. The ratio of two isotopes of the same element may change in a mineral as one of them decays or is produced radioactively, and the isotopes of some of the lighter elements, such as carbon and oxygen, may change during chemical reactions like the growth of organic matter or the evaporation of seawater. However, most chemical reactions do not shift the isotopic ratios of the elements involved. The ratio thus provides a kind of label on an element, allowing geochemists to follow its pathways through the maze of chemical transformations in which minerals dissolve, precipitate, and react. The study of isotopes has contributed enormously to our understanding of how the Earth works.

20. DePaolo, D. J., Kyte, F. T., Marshall, B. D., O'Neil, J. R., and Smit, J., 1983, Rb-Sr, Sm-Nd, K-Ca, O, and H isotopic study of Cretaceous-Tertiary boundary sediments, Caravaca, Spain: evidence for an oceanic impact site: Earth and Planetary Science Letters, v. 64, p. 356–373.

21. Montanari, A., Hay, R. L., Alvarez, W., Asaro, F., Michel, H. V., Alvarez, L. W., and Smit, J., 1983, Spheroids at the Cretaceous-Tertiary boundary are altered impact droplets of basaltic composition: Geology, v. 11, p. 668–671.

22. Bohor, B. F., Foord, E. E., Modreski, P. J., and Triplehorn, D. M., 1984, Mineralogic evidence for an impact event at the Cretaceous-Tertiary boundary: Science, v. 224, p. 867–869.

23. French, B. M. and Short, N. M., eds., 1968, Shock metamorphism of natural materials: Baltimore, Mono Book Corp., 644 p.

24. Izett, G. A., 1990, The Cretaceous/Tertiary boundary interval, Raton Basin, Colorado and New Mexico, and its content of shock-metamorphosed minerals; evidence relevant to the K/T boundary impact-extinction theory: Geological Society of America Special Paper, v. 249, 100 p. Stishovite, a high-pressure form of quartz, and absolutely diagnostic of impact shock was reported from the Raton Basin KT

boundary: McHone, J. F., Nieman, R. A., Lewis, C. F., and Yates, A. M., 1989, Stishovite at the Cretaceous-Tertiary boundary, Raton, New Mexico, Science, v. 243, p. 1182–1184. In addition, Michael Owen and Mark Anders showed that the KT shocked-quart grains gave cathodoluminescence colors very different from those of volcanic quartz grains: Owen, M. R. and Anders, M. H., 1988, Evidence from cathodoluminescence for non-volcanic origin of shocked quartz at the Cretaceous-Tertiary boundary: Nature, v. 334, p. 145–147.

25. Carter, N. L., Officer, C. B., and Drake, C. L., 1990, Dynamic deformation of quartz and feldspar: clues to causes of some natural crises: Tectonophysics, v. 171, p. 373–391.

26. Vogt, P. R., 1972, Evidence for global synchronism in mantle plume convection, and possible significance for geology: Nature, v. 240, p. 338–342; McLean, D. M., 1982, Deccan volcanism and the Cretaceous-Tertiary transition scenario: a unifying casual mechanism: Syllogeus, v. 39, p. 143–144.

27. Courtillot, V., Besse, J., Vandamme, D., Montigny, R., Jaeger, J.-J., and Cappetta, H., 1986, Deccan flood basalts at the Cretaceous/Tertiary boundary?: Earth and Planetary Science Letters, v. 80, p. 361–374; Courtillot, V. E., 1990, What caused the mass extinction? A volcanic eruption: Scientific American, v. 263 (October), p. 85–92.

28. Officer, C. B. and Drake, C. L., 1983, The Cretaceous-Tertiary transition: Science, v. 219, p. 1383–1390; Officer, C. B. and Drake, C. L., 1985, Terminal Cretaceous environmental effects: Science, v. 227, p. 1161–1167.

29. Raup, D. M. and Sepkoski, J. J., Jr., 1984, Periodicity of extinctions in the geologic past: Proceedings of the National Academy of Sciences, v. 81, p. 801–805.

30. Davis, M., Hut, P., and Muller, R. A., 1984, Extinction of species by periodic comet showers: Nature, v. 308, p. 715–717.

31. Alvarez, W. and Muller, R. A., 1984, Evidence from crater ages for periodic impacts on the Earth: Nature, v. 308, p. 718–720.

32. Whitmire, D. P. and Jackson, A. A., IV, 1984, Are periodic mass extinctions driven by a distant solar companion?: Nature, v. 308, p. 713–715; Rampino, M. R. and Stothers, R. B., 1984, Terrestrial mass extinctions, cometary impacts and the Sun's motion perpendicular to the galactic plane: Nature, v. 308, p. 709–712; Whitmire, D. P. and Matese, J. J., 1985, Periodic comet showers and Planet X: Nature, v. 313, p. 36–38.

33. Muller, R. A., 1988, Nemesis: the death star (The story of a scientific revolution): New York, Weidenfeld and Nicolson, 193 p. For another account, see Raup, D. M., 1986, The Nemesis affair: New York, W. W. Norton, 220 p.

34. Zoller, W. H., Parrington, J. R., and Phelan Kotra, J. M., 1983, Iridium enrichment in airborne particles from Kilauea Volcano: January 1983: Science, v. 222, p. 1118–1121; Olmez, I., Finnegan, D. L., and Zoller, W. H., 1986, Iridium emissions from Kilauea Volcano: Journal of Geophysical Research, v. 91, p. 653–663.

35. Bekov, G. I., Letokhov, V. S., Radaev, V. N., Badyukov, D. D., and Nazarov, M. A., 1988, Rhodium distribution at the Cretaceous/Tertiary boundary analysed by ultrasensitive laser photoionization: Nature, v. 332, p. 146–148.

36. Alvarez, W., Asaro, F., and Montanari, A., 1990, Iridium profile for 10 million years across the Cretaceous-Tertiary boundary at Gubbio (Italy): Science, v. 250, p. 1700–1702.

37. In a special ceremony at Lawrence Berkeley Laboratory in October, 1995, Frank formally christened the instrument the "Luis W. Alvarez Iridium Coincidence Spectrometer."

38. Muller, R. A., 1985, Evidence for a solar companion star, in Papagiannis, M. D., ed., The search for extraterrestrial life: recent developments: Dordrecht, Riedel, p. 233–243.

39. Hut, P., Alvarez, W., Elder, W. P., Hansen, T., Kauffman, E. G., Keller, G., Shoemaker, E. M., and Weissman, P. R., 1987, Comet showers as a cause of mass extinctions: Nature, v. 329, p. 118–126; Montanari, A., 1990, Geochronology of the terminal Eocene impacts; an update: Geological Society of America Special Paper, v. 247, p. 607–616; Montanari, A., Asaro, F., and Kennett, J. P., 1993, Iridium anomalies of late Eocene age at Massignano (Italy), and ODP Site 689B (Maud Rise, Antarctica): Palaios, v. 8, p. 420–437.

40. Izett, G. A., 1990, op. cit.

41. Hartung, J. B. and Anderson, R. R., 1988, A compilation of information and data on the Manson impact structure: Houston, Lunar and Planetary Institute, 32 p.

42. Anderson, R. R., Hartung, J. B., Shoemaker, E. M., and Roddy, D. J., 1991, New research core drilling in the Manson impact structure, Iowa: A first look at the spectacular rocks formed at a K-T boundary impact site: Geological Society of America Abstracts with Programs, v. 23, p. A402; Koeberl, C. and Anderson, R. R., eds., 1996, The Manson

impact structure, Iowa: anatomy of an impact crater: Geological Society of America Special Paper, v. 302, p. 468 Steiner, M. B., 1996, Implications of magneto-mineralogic characteristics of the Manson and Chicxulub impact rocks, in Ryder, G., Fastovsky, D., and Gartner, S., eds., The Cretaceous-Tertiary event and other catastrophes in Earth history: Geological Society of America Special Paper, v. 307, p. 89–104.

43. Izett, G. A., Cobban, W. A., Obradovich, J. D., and Kunk, M. J., 1993, The Manson impact structure: 40Ar/39Ar age and its distal impact ejecta in the Pierre Shale in southeastern South Dakota: Science, v. 262, p. 729–732.

CHAPTER SIX
THE CRATER OF DOOM

1. Maurrasse, F. J.-M. R., 1980, New data on the stratigraphy of the Southern Peninsula of Haiti, in Maurrasse, F. J.-M. R., ed., Présentations Transactions du Colloque sur la Géologie d'Haïti: Port-au-Prince, p. 184–198.

2. Alvarez, W., Alvarez, L. W., Asaro, F., and Michel, H. V., 1982, Current status of the impact theory for the terminal Cretaceous extinction: Geological Society of America Special Paper, v. 190, p. 305–315.

3. Hansen, T., 1982, Macrofauna of the Cretaceous/Tertiary boundary interval in east-central Texas, in Maddocks, R. F., ed., Texas Ostracoda: Houston, Department of Geosciences, University of Houston, p. 231–237.

4. Smit, J. and Romein, A. J. T., 1985, A sequence of events across the Cretaceous-Tertiary boundary: Earth and Planetary Science Letters, v. 74, p. 155–170.

5. Bourgeois, J., Hansen, T. A., Wiberg, P. L., and Kauffman, E. G., 1988, A tsunami deposit at the Cretaceous-Tertiary boundary in Texas: Science, v. 241, p. 567–570.

6. Hildebrand, A. R., Boynton, W. V., and Zoller, W. H., 1984, Kilauea volcano aerosols: evidence in siderophile element abundances for impact-induced oceanic volcanism at the K/T boundary: Meteoritics, v. 19, p. 239–240; Hildebrand, A. R. and Boynton, W. V., 1987, The K/T impact excavated oceanic mantle: evidence from REE abun-dances: Lunar and Planetary Science, v. 18, p. 427–428.

7. Hildebrand, A. R., Penfield, G. T., Kring, D. A., Pilkington, M., Camargo Z., A., Jacobsen, S. B., and Boynton, W. V., 1991, Chicxulub

crater: a possible Cretaceous/Tertiary boundary impact crater on the Yucatán Peninsula, Mexico: Geology, v. 19, p. 867–871.

8. Cornejo T., A. and Hernandez O., A., 1950, Las anomalías gravimétricas en la Cuenca Salina del Istmo, Planicie Costera de Tabasco, Campeche y Península de Yucatán: Boletín de la Asociación Mexicana de Geólogos Petroleros, v. 2 p. 453–460.

9. Penfield, G. T. and Camargo Z., A., 1981, Definition of a major igneous zone in the central Yucatán platform with aeromagnetics and gravity: Society of Exploration Geophysicists Technical Program, Abstracts, and Biographies, v. 51, p. 37.

10. Muir, J. M., 1936, Geology of the Tampico region, Mexico: Tulsa, American Association of Petroleum Geologists, 280 p.

11. Gamper, M. A., 1977, Acerca del límite Cretácio-Tericiario en México: Universidad Nacional Autónoma de México, Instituto de Geología, Revista, v. 1, p. 23–27; Gamper, M. A., 1977, Bioestratigrafia del Paleoceno y Eoceno de la Cuenca Tampico-Mislanta basada en los foraminíferos planctónicos: Universidad National Autónoma de México, Instituto de Geología, Revista, v. 1, p. 117–128; Gamper-Longoria, M. A. and Longoria, J., 1984, Foraminiferal biochronology at the Cretaceous/Tertiary boundary: Geological Society of America Abstracts with Programs, v. 16, p. 84.

12. Sigurdsson, H., D'Hondt, S., Arthur, M. A., Bralower, T. J., Zachos, J. C., Van Fossen, M., and Channell, J. E. T., 1991, Glass from the Cretaceous-Tertiary boundary in Haiti: Nature, v. 349, p. 482–487; Izett, G. A., 1991, Tektites in Cretaceous-Tertiary boundary rocks on Haiti and their bearing on the Alvarez impact extinction hypothesis: Journal of Geophysical Research, v. 96, p. 20,879–20,905; Maurrasse, F. J.-M. R. and Sen, G., 1991, Impacts, tsunamis, and the Haitian Cretaceous-Tertiary boundary layer: Science, v. 252, p. 1690–1693; Lyons, J. B. and Officer, C. B., 1992, Mineralogy and petrology of the Haiti Cretaceous-Tertiary section: Earth and Planetary Science Letters, v. 109, p. 205–224.

13. The first microtektites, from a much younger impact, had been discovered by Billy Glass: Glass, B. P., 1967, Microtektites in deep-sea sediments: Nature, v. 214, p. 372–374.

14. Sigurdsson, H., Bonté, P., Turpin, L., Chaussidon, M., Metrich, N., Steinberg, M., Pradel, P., and D'Hondt, S., 1991, Geochemical constraints on source region of Cretaceous/Tertiary impact glasses: Nature, v. 353, p. 839–842.

15. Margolis, S. V., Claeys, P., and Kyte, F., T., 1991, Microtektites,

microkrystites and spinels from a Late Piocene asteroid impact in the southern ocean: Science, v. 251 p. 1594–1597.

16. Redrilling the crater using modern techniques has become a major goal of scientists interested in the KT boundary event. Buck Sharpton at the Lunar and Planetary Institute in Houston and Luis Marín at UNAM, the National University in Mexico City, are leading the effort to get this work done.

17. Swisher, C. C., III, Grajales-Nishimura, J. M., Montanari, A., Margolis, S. V., Claeys, P., Alvarez, W., Renne, P., Cedillo-Pardo, E., Maurrasse, F. J.-M. R., Curtis, G. H., Smit, J., and McWilliams, M. O., 1992, Coeval 40Ar/39Ar ages of 65.0 million years ago from Chicxulub Crater melt rock and Cretaceous-Tertiary boundary tektites: Science, v. 257, p. 954–958; Sharpton, V. L., Dalrymple, G. B., Marín, L. E., Ryder, G., Schuraytz, B. C., and Urrutia, J., 1992, New links between the Chicxulub impact structure and the Cretaceous/Tertiary boundary: Nature, v. 359, p. 819–821.

18. Blum, J. D. and Chamberlain, C. P., 1992, Oxygen isotope constraints on the origin of impact glasses from the Cretaceous-Tertiary boundary: Science, v. 257, p. 1104–1107; Blum, J. D., Chamberlain, C. P., Hingston, M. P., Koeberl, C., Marin, L. E., Schuraytz, B. C., and Sharpton, V. L., 1993, Isotopic comparison of K/T boundary impact glasses with melt rock from the Chicxulub and Manson impact structures: Nature, v. 364, p. 325–327.

19. Smit, J., Montanari, A., Swinburne, N. H. M., Alvarez, W., Hildebrand, A. R., Margolis, S. V., Claeys, P., Lowrie, W., and Asaro, F., 1992, Tektite-bearing, deep-water clastic unit at the Cretaceous-Tertiary boundary in northeastern Mexico: Geology, v. 20, p. 99–103; Alvarez, W., Smit, J., Lowrie, W., Asaro, F., Margolis, S. V., Claeys, P., Kastner, M., and Hildebrand, A. R., 1992, Proximal impact deposits at the Cretaceous-Tertiary boundary in the Gulf of Mexico: A restudy of DSDP Leg 77 Sites 536 and 540: Geology, v. 20, p. 697–700.

20. Officer, C. B., Drake, C. L., Pindell, J. L., and Meyerhoff, A. A., 1992, Cretaceous-Tertiary events and the Caribbean caper: GSA Today, v. 2, p. 69–75; Officer, C. B. and Page, J., 1996, The great dinosaur extinction controversy: Reading, MA, Addison-Wesley, 209 p.

21. Keller, G., MacLeod, N., Lyons, J. B., and Officer, C. B., 1993, Is there evidence for Cretaceous-Tertiary boundary-age deep-water deposits in the Caribbean and Gulf of Mexico?: Geology, v. 21, p. 776–780. For criticisms of this paper and the responses of Keller and her colleagues, see Geology, v. 22, p. 953–958, 1993.

22. Alvarez, W., Grajales N., J. M., Martinez S., R., Romero M., P. R., Ruiz L., E., Guzmán R., M. J., Zambrano A., M., Smit, J., Swinburne, N. H. M., and Margolis, S. V., 1992, The Cretaceous-Tertiary boundary impact-tsunami deposit in NE Mexico: Geological Society of America Abstracts with Programs, v. 24, p. A331.

23. There were other lines of evidence supporting impact, which I have not discussed in detail in the text of the book: (1) The KT boundary spherules contain nickel-rich spinels crystallized from the vapor cloud and derived in part from the impactor (Smit, J. and Kyte, F. T., 1994, Siderophile-rich magnetic spheroids from the Cretaceous-Tertiary boundary in Umbria, Italy: Nature, v. 310, p. 403–405; Bohor, B. F., Foord, E. E., and Ganapathy, R., 1986, Magnesioferrite from the Cretaceous-Tertiary boundary, Caravaca, Spain: Earth and Planetary Science Letters, v. 81, p. 57–66; Kyte, F. T. and Smit, J., 1986, Regional variations in spinel compositions: An important key to the Cretaceous/Tertiary event: Geology, v. 14, p. 485–487; Robin, E., Boclet, D., Bonté, P., Froget, L., Jéhanno, C., and Rocchia, R., 1991, The stratigraphic distribution of Ni-rich spinels in Cretaceous-Tertiary boundary rocks at El-Kef (Tunisia), Caravaca (Spain), and Hole-761C (Leg-122): Earth Plan, v. 107, p. 715–721; Robin, E., Bonté, P., Froget, L., Jéhanno, C., and Rocchia, R., 1992, Formation of spinels in cosmic objects during atmospheric entry: a clue to the Cretaceous-Tertiary boundary event: Earth and Planetary Science Letters, v. 108, p. 181–190). (2) Extraterrestrial amino acids were recovered from the KT boundary clay at Stevns Klint (Zhao, M. and Bada, J. L., 1989, Extraterrestrial amino acids in Cretaceous/Tertiary boundary sediments at Stevns Klint, Denmark: Nature, v. 339, p. 463–465). (3) Tiny impact diamonds have been recovered from the KT boundary (Carlisle, D. B., 1992, Diamonds at the K/T boundary: Nature, v. 357, p. 119–120; Carlisle, D. B. and Braman, D. R., 1991, Nanometre-size diamonds in the Cretaceous Tertiary boundary clay of Alberta: Nature, v. 352, p. 708–709; Gilmour, I., Russell, S. S., Arden, J. W., Lee, M. R., Franchi, I. A., and Pillinger, C. T., 1992, Terrestrial carbon and nitrogen isotopic ratios from Cretaceous-Tertiary boundary nanodiamonds: Science, v. 258, p. 1624–1626). (4) The isotopic ratio of osmium showed this platinum-group element in the KT boundary to have come from a meteoric source (Luck, J.-M. and Turekian, K. K., 1983, Osmium-187/Osmium-186 in manganese nodules and the Cretaceous-Tertiary boundary: Science, v. 222, p. 613–615).

24. Smit, J., Roep, Th. B., Alvarez, W., Montanari, A., Claeys, P.,

Grajales-Nishimura, J. M., and Bermudez, J., 1996, Coarse-grained, clastic sandstone complex at the K/T boundary around the Gulf of Mexico: deposition by tsunami waves induced by the Chicxulub impact?, in Ryder, G., Fastovsky, D., and Gartner, S., eds., The Cretaceous-Tertiary event and other catastrophes in Earth history: Geological Society of America Special Paper, v. 307, p. 151–182.

CHAPTER SEVEN
THE WORLD AFTER CHICXULUB

1. Sharpton, V. L., Burke, K., Camargo Zanoguera, A., Hall, S. A., Lee, D. S., Marín, L. E., Suárez Reynoso, G., Quezada Muñeton, J. M., Spudis, P. D., and Urrutia Fucugauchi, J., 1993, Chicxulub multiring impact basin: size and other characteristics derived from gravity analysis: Science, v. 261, p. 1564–1567.

2. Hildebrand, A. R., Pilkington, M., Connors, M., Ortiz Aleman, C., and Chavez, R. E., 1995, Size and structure of the Chicxulub crater revealed by horizontal gravity gradients and cenotes: Nature, v. 376, p. 415–417.

3. Camargo Z., A. and Suárez R., G., 1994, Evidencia sísmica del cráter de impacto de Chicxulub: Boletín de la Asociación Mexicana de Geofísicos de Exploración, v. 34, p. 1–28.

4. Pope, K. O., Ocampo, A. C., and Duller, C. E., 1991, Mexican site for K/T impact crater: Nature, v. 351, p. 105; Pope, K. O., Ocampo, A. C., and Duller, C. E., 1993, Surficial geology of the Chicxulub impact crater, Yucatán, Mexico: Earth, Moon, and Planets, v. 63, p. 93–104.

5. Alvarez, W., 1996, Trajectories of ballistic ejecta from the Chicxulub Crater, in Ryder, G., Fastovsky, D., and Gartner, S., eds., The Cretaceous-Tertiary event and other catastrophes in Earth history: Geological Society of America Special Paper, v. 307, p. 141–150.

6. Bostwick, J. A. and Kyte, F. T., 1996, The size and abundance of shocked quartz in Cretaceous-Tertiary boundary sediments from the Pacific Basin, in Ryder, G., Fastovksy, D., and Gartner, S., eds., The Cretaceous-Tertiary event and other catastrophes in Earth history: Geological Society of America Special Paper, v. 307, p. 403–415.

7. Bohor, B. F., 1990, Shocked quartz and more; Impact signatures in Cretaceous/Tertiary boundary clays, in Sharpton, V. L. and Ward, P. D., eds., Global catastrophes in Earth history: Geological Society of

America Special Paper, v. 247, p. 335–342; Izett, G. A., 1990, The Cretaceous/Tertiary boundary interval, Raton Basin, Colorado and New Mexico, and its content of shock-metamorphosed minerals; evidence relevant to the K/T boundary impact-extinction theory: Geological Society of America Special Paper, v. 249, p. 1–100.

8. Kieffer, S. W., 1981, Fluid dynamics of the May 18 blast at Mount St. Helens: U.S. Geological Survey Professional Paper, v. 1250, p. 379–400; Kieffer, S. W., 1989, Geologic nozzles: Reviews of Geophysics, v. 21, p. 3–38; Simonds, C. H. and Kieffer, S. W., 1993, Impact and volcanism; a momentum scaling law for erosion: Journal of Geophysical Research, B, v. 98, p. 14,321–14,337.

9. Melosh, H. J., 1988, Impact cratering: a geologic process: Tucson, University of Arizona Press, 272 p.

10. Alvarez, W., Claeys, P., and Kieffer, S.W., 1995, Emplacement of KT boundary shocked quartz from Chicxulub crater: Science, v. 269, p. 930–935.

11. The latest compilation of impact craters is by Grieve, R. A. F. and Pesonen, L. J., 1992, The terrestrial impact cratering record: Tectonophysics, v. 216, p. 1–30.

12. Lowe, D. R. and Byerly, G. R., 1986, Early Archean silicate spherules of probable impact origin, South Africa and western Australia: Geology, v. 14, p. 83–86; Lowe, D. R., Byerly, G. R., Asaro, F., and Kyte, F. T., 1989, Geological and geochemical record of 3,400-million-year-old terrestrial meteorite impacts: Science, v. 245, p. 959–962.

13. Gostin, V. A., Haines, P. W., Jenkins, R. J. F., Compston, W., and Williams, I. S., 1986, Impact ejecta horizon within Late Precambrian shales, Adelaide Geosyncline, South Australia: Science, v. 233, p. 198–200; Williams, G. E., 1986, The Acraman impact structure: source of ejecta in Late Precambrian shales, South Australia: Science, v. 233, p. 200–203; Gostin, V. A., Keays, R. R., and Wallace, M. W., 1989, Iridium anomaly from the Acraman impact ejecta horizon: impacts can produce sedimentary iridium peaks: Nature, v. 340, p. 542–544; Wallace, M. W., Gostin, V. A., and Keays, R. R., 1990, Acraman impact ejecta and host shales: Evidence for low-temperature mobilization of iridium and other platinoids: Geology, v. 18, p. 132–135.

14. Raup, D. M., 1991, Extinction—Bad genes or bad luck?: New York, W. W. Norton, 210 p.

15. McLaren, D. J., 1970, Presidential address: Time, life and boundaries: Journal of Paleontology, v. 44, p. 801–815.

16. Wang, K., Orth, C. J., Attrep, M., Jr., Chatterton, B. D. E., Hou, H., and Geldsetzer, H. H. J., 1991, Geochemical evidence for a catastrophic biotic event at the Frasnian/Fammenian boundary in south China: Geology, v. 19, p. 776–779; Wang, K., 1992, Glassy microspherules (microtektites) from an Upper Devonian limestone: Science, v. 256, p. 1547–1550.

17. Claeys, P., Casier, J.-G., and Margolis, S. V., 1992, Microtektites and mass extinctions: evidence for a Late Devonian asteroid impact: Science, v. 257, p. 1102–1104.

18. Bice, D. M., Newton, C. R., McCauley, S., Reiners, P. W., and McRoberts, C. A., 1992, Shocked quartz at the Triassic-Jurassic boundary in Italy: Science, v. 255, p. 443–446.

19. Montanari, A., Asaro, F., and Kennett, J. P., 1993, Iridium anomalies of late Eocene age at Massignano (Italy), and ODP Site 689B (Maud Rise, Antarctica); Palaios, v. 8, p. 420–437; Clymer, A. K., Bice, D. M., and Montanari, A., 1996, Shocked quartz from the late Eocene: impact evidence from Massignano, Italy: Geology, v. 24, p. 483–486.

20. Masaitis, V. L., 1994, Impactites from Popigai crater, in Dressler, B. O., Grieve, R. A. F., and Sharpton, V. L., eds., Large meteorite impacts and planetary evolution: Geological Society of America Special Paper, v. 293, p. 153–162; Koeberl, C., Poag, C. W., Reimold, W. U., and Brandt, D., 1996, Impact origin of the Chesapeake Bay structure and the source of the North American tektites: Science, v. 271, p. 1263–1266.

21. Kyte, F. T., Zhou, Z., and Wasson, J. T., 1981, High noble metal concentrations in a late Pliocene sediment: Nature, v. 292, p. 417–420; Kyte, F. T., Zhou, Z., and Wasson, J. T., 1988, New evidence on the size and possible effects of a Late Pliocene oceanic asteroid impact: Science, v. 241, p. 63–65; Margolis, S. V., Claeys, P., and Kyte, F. T., 1991, Microtektites, microkrystites and spinels from a Late Pliocene asteroid impact in the southern ocean: Science, v. 251, p. 1594–1597.

22. Leroux, H., Warme, J. E., and Doukhan, J.-C., 1995, Shocked quartz in the Alamo breccia, southern Nevada: Evidence for a Devonian impact event: Geology, v. 23, p. 1003–1006.

23. Warme, J. E. and Sandberg, C. A., 1996, Alamo megabreccia: Record of a Late Devonian impact in southern Nevada: GSA Today, v. 6, p. 1–7.

24. "Traps" comes from a Dutch word for "steps," and refers to the stair-step topography produced by erosion of the Deccan basalt flows

in India. By extension, "traps" has been applied to other large basalt provinces on the continents. The largest of these is the Siberian Traps.

25. Erwin, D. H., 1993, The great Paleozoic crisis: life and death in the Permian: New York, Columbia University Press, 327 p.; Erwin, D. H., 1994, The Permo-Triassic extinction: Nature, v. 367, p. 231–236; Erwin, D. H., 1996, The mother of mass extinctions: Scientific American, v. 275 (July), p. 72–78.

26. Renne, P. R., Zhang, Z., Richards, M. A., Black, M. T., and Basu, A. R., 1995, Synchrony and causal relations between Permian-Triassic boundary crises and Siberian flood volcanism: Science, v. 269, p. 1413–1416.

27. It has been suggested that impact at one location triggers volcanism where seismic energy converges at the point exactly on the opposite side of the globe, but at KT boundary time, India was 3,000 km away from the point antipodal to Chicxulub, so this suggestion does not seem to work.

28. Levy, D. H., 1995, Impact Jupiter: the crash of comet Shoemaker-Levy 9: New York, Plenum Press, 290 p.; Spencer, J. R. and Mitton, J., eds., 1995, The great comet crash: the impact of comet Shoemaker-Levy 9 on Jupiter: New York, Cambridge University Press, 118 p; Dauber, P. M. and Muller, R. A., 1996, The three big bangs: New York, Addison-Wesley, 207 p., ch. 2.

Italicized page numbers refer to photographs and figures

172